CIVIL ENGINEERING CONTRACT ADMINISTRATION AND CONTROL

CIVIL ENGINEERING CONTRACT ADMINISTRATION AND CONTROL

IVOR H. SEELEY

BSc, MA, PhD, CEng, FICE, FRICS, MCIOB

*Emeritus Professor of Trent Polytechnic, Nottingham
and Chartered Civil Engineer and Surveyor*

MACMILLAN

© Ivor H. Seeley 1986

First published 1986.

Published by
MACMILLAN EDUCATION LTD
Houndmills, Basingstoke, Hampshire RG21 2XS
and London
Companies and representatives
throughout the world

Printed in Hong Kong

British Library Cataloguing in Publication Data
Seeley, Ivor H.
    Civil engineering contract administration and
    control.—(Macmillan building and surveying
    series)
    1. Civil engineering—Contracts and specifications
    —Great Britain
    I. Title
    642   TA181.G7

ISBN 0-333-40591-9
ISBN 0-333-40592-7 Pbk

Series Standing Order

If you would like to receive future titles in this series as they are
published, you can make use of our standing order facility. To place a
standing order please contact your bookseller or, in case of difficulty,
write to us at the address below with your name and address and the
name of the series. Please state with which title you wish to begin your
standing order. (If you live outside the United Kingdom we may not
have the rights for your area, in which case we will forward your order
to the publisher concerned.)

Customer Services Department, Macmillan Distribution Ltd
Houndmills, Basingstoke, Hampshire, RG21 2XS, England.

'For which of you, intending to build a tower, sitteth not down first, and counteth the cost, whether he have sufficient to finish it?'

Luke 14.28

Contents

Preface

This book is aimed at civil engineers and civil engineering contractors and at students on civil engineering degree and diploma courses. Allied professions such as construction and quantity surveyors may find its contents of interest and value.

The diverse aspects of civil engineering contract administration are examined and described, with relevant supporting examples. It starts by considering the general backcloth to civil engineering works and contracts, including funding, preliminary investigations and the preparation of engineer's reports. The form and purpose of the various contract documents are examined and the principal requirements of the ICE Conditions summarised and explained.

The principal tendering arrangements are described and compared, together with the more commonly practised approaches to estimating the cost of civil engineering works.

Site organisation and supervision are covered in sufficient depth to illustrate the means by which a civil engineering project can be effectively planned, managed and controlled, and having regard to such important aspects as productivity, plant usage and safety of operatives.

One of the major problems encountered in the operation of civil engineering contracts is a weakness in communication between the parties to the contract and site personnel. Vital communication requirements are described and illustrated, including the preparation of site documents and records and the purpose and conduct of site meetings.

The method of measuring and valuing civil engineering works is explored and this encompasses the use of daywork, issue of interim certificates, settlement of final accounts, valuation of variations and financial control of contracts.

Finally, the book examines the background to contractors' claims and how they should be presented by the contractor and dealt with by the engineer. The disruptive effect of variations and the main causes of loss and expense to the contractor are considered at length, together with the assessment of liquidated damages.

Nottingham   IVOR H. SEELEY
Spring 1986

Acknowledgements

A debt of gratitude is owed to many civil engineers who have over the years given me the benefit of their knowledge and experience.

My thanks are due to the Institution of Civil Engineers for kind permission to quote from the *Conditions of Contract for use in connection with Works of Civil Engineering Construction* and to reproduce figure 4.1 from *Civil Engineering Procedure* published by Thomas Telford Ltd. Permission was also given to reproduce figure 4.2 from *The Practice of Construction Management* by Barry Fryer, 1985, Collins Professional and Technical Books.

Macmillan Publishers kindly gave permission to quote from *Civil Engineering Quantities*, *Civil Engineering Specification* and *Quantity Surveying Practice*.

I received abundant help and consideration from the publishers throughout the preparation and production stages of the book. Mrs. E.D. Robinson prepared the typescript quickly and efficiently and my wife played her customary supportive role.

List of Tables

List of Figures

1 Contractual Arrangements

This chapter examines the wide range of civil engineering works, the nature and functions of the parties to civil engineering contracts, the form and use of the different contractual arrangements, preliminary investigations, project reports and the selection of contractors.

SCOPE OF CIVIL ENGINEERING WORKS

The Royal Charter of the Institution of Civil Engineers[1] defined the profession of a civil engineer as 'the art of directing the great sources of power in Nature for the use and convenience of man, as the means of production and of traffic in states both for external and internal trade, as applied in the construction of roads, bridges, aqueducts, canals, river navigation and docks, for internal intercourse and exchange, and in the construction of ports, harbours, moles, breakwaters and lighthouses, and in the art of navigation by artificial power for the purposes of commerce, and in the construction and adaptation of machinery, and in the drainage of cities and towns.'

Civil engineering works encompass a wide range of different projects, some of which are of great magnitude, such as the Hong Kong £3000m mass transit railway system, the Singapore £2000m rapid transit system, the Changi Airport at Singapore and the £200m fourth terminal at London's Heathrow airport. Vast cuttings and embankments; large mass and reinforced concrete structures, such as the frameworks of large buildings, reservoirs, sea walls, bridges and cooling towers for power stations; structural steel frameworks of large buildings; piling for heavy foundations; jetties, wharves and dry docks; long pipe-lines, tunnels and railway trackwork, all comprise civil engineering projects.

Civil engineering work also embraces structural engineering projects constructed in various materials, and public works engineering such as roads, bridges, sewers, sewage treatment works, water mains, reservoirs, water towers, works of river and sea defence, refuse disposal plants, marinas and swimming pools, carried out on behalf of local and water authorities.[2]

These works require considerable technical knowledge, skill and ingenuity in both their design and construction. The use of new materials, plant and techniques is continually changing the nature and methods of construction used in these projects, and the increasing size and complexity of these works call for ever greater knowledge, skill and expertise on the part of engineers and contractors alike.

Furthermore, the Institution of Civil Engineers[3] has described how, as countries become more developed, social costs and effects on the environment assume greater importance, and the standards by which the social consequences of civil engineering are judged become more exacting. The engineer must, as far as is practicable, take these factors into account in formulating his designs. Civil engineering solutions should not be divorced from social, economic and ecological effects, working in liaison with other disciplines, where necessary, and keeping the employer fully informed.

THE INSTITUTION OF CIVIL ENGINEERS

The Institution of Civil Engineers is the principal professional body encompassing the whole range of civil engineering works. It maintains high professional standards by means of its exacting education, training and membership requirements, professional development activities and the operation of rigorous rules of professional conduct. The membership of the Institution encompasses civil engineers in private and public offices and in contracting organisations. Unification of the Institution of Municipal Engineers with the Institution of Civil Engineers took place in April 1984, as they both had many common interests and together form a stronger body with wider interests and greater impact. An Association of Municipal Engineers was formed within the ICE to provide a focus for those involved in municipal engineering and the Association publishes a bi-monthly journal. The main ICE Journal *New Civil Engineer* is issued weekly.

The Institution has produced the *ICE Conditions of Contract* in collaboration with the Association of Consulting Engineers and the Federation of Civil Engineering Contractors. These Conditions are examined in chapter 2. The Arbitration Advisory Board is responsible for all aspects of arbitration which may arise from civil engineering contracts. In the area of public health and safety, the Institution has a statutory function under the Reservoirs (Safety Provisions) Act 1930 and the Reservoirs Act 1975. It also plays an important role in the scheme for training civil engineering technicians and technician engineers. The ICE Panel for Historical Works identifies important historic works which form part of the national heritage and takes positive action to preserve, record and publicise them.

FUNDING OF PROJECTS

Promoters of projects have to find the necessary finance, be they public or private. Contracts undertaken by government departments are principally defence installations, roads, bridges and office buildings, forming part of the national infrastructure. The funds are derived from taxation and other forms of government revenue, and expenditure is monitored by the relevant government departments and the Treasury.

Local authorities undertake construction work for the benefit of the residents in their areas, and also wider-based schemes as agents of central government. The finance comes mainly from local rates, government grants and loans. In the early nineteen-eighties, local authority expenditure on construction work was reduced significantly by cuts in government funding and rate-capping.

Civil engineering projects are also undertaken for other public bodies such as Water Authorities and the Central Electricity Generating Board who obtain their funds primarily through consumers' payments for the services provided. New town development corporations have responsibility for the construction of new towns and receive their initial funds from the Treasury but are subsequently expected to finance their own operations through revenue and the sale of assets, and are subject to government financial control.

On the wider front, construction projects may be assisted financially by such organisations as the World Bank, the International Monetary Fund and the Asian Development Bank. The financial assistance provided by these bodies is mainly in the form of loans made to developing countries.

Registered companies promote a wide range of civil engineering projects in the private sector. The companies can raise capital by public subscription for their shares. Boards of directors exercise internal financial control and companies' accounts are audited annually.

Construction companies obtain their funds from a variety of sources. For example, in 1978 the larger construction companies obtained approximately 75 per cent of their funds from shareholders' interests, 5 per cent from debentures and loans and the remaining 20 per cent from banks and short loans.

PARTIES TO A CONTRACT

The principal parties to a civil engineering contract are the employer or promoter, and the contractor. The contract is for a specific project, as detailed in the contract documents, to be carried out by the contractor in return for payment from the employer.

The employer is normally a corporate body such as a government department, local authority, statutory body or limited liability company. The term 'employer' is rather misleading as his relationship with an independent contractor is entirely different from the normal employer/employee arrangement encountered in most industrial and commercial concerns. Hence ICE publications refer to the employer as the promoter.[4]

The employer appoints an engineer as his representative to design the project, supervise the constructional work, arrange for payments to the contractor and settle any disputes. Most public authorities employ their own engineers to take charge of many of the authorities' construction contracts, although firms of consulting civil engineers are given responsibility for many major civil engineering schemes.

The bulk of civil engineering work is carried out by contractors who are usually limited liability companies. Some of the larger contractors undertake both building and civil engineering work, and many civil engineering contracts contain some building work and the demarcation between the two classes of work is, on occasions, rather blurred. Some contractors specialise in particular classes of work such as demolition, earthworks, piling, tunnelling and marine work.

The Federation of Civil Engineering Contractors was established in 1919 to protect the interests of its members. Other aims are to establish satisfactory relations between members and their employees, to regulate wages and working conditions in the industry, to maintain a high standard of conduct, to combat unfair practices and encourage efficiency among it members, and to settle and secure the adoption in civil engineering contracts of a standard form of contract embodying equitable conditions.

CIVIL ENGINEERING CONTRACTS

Nature and Form of Contracts

The law relating to civil engineering contracts is one aspect of the law relating to contract and tort (civil wrongs). A simple contract consists of an agreement entered into by two or more parties, whereby one of the parties undertakes to do something in return for something to be undertaken by the other. A contract has been defined as an agreement which directly creates and contemplates an obligation. In general, English law requires no special formalities in making contracts but, for various reasons, some contracts must be made in a particular form to be enforceable and, if they are not made in that special way, then they will be ineffective.

Some contracts must be made 'under seal', for example, Deeds of Gift or any contract where 'consideration' is not present (consideration is defined later in the chapter). Some other contracts must be in writing, for example, one covering the Assignment of Copyright, where an Act of Parliament specifically states that writing is necessary. Since the passing of the Corporate Bodies Contracts Act 1960, contracts entered into by corporations, including local authorities, can be binding without being made under seal.

It is sufficient, in order to create a legally binding contract, if the parties express their agreement and intention to enter into such a contract. If, however, there is no written agreement and a dispute arises in respect of the contract, then the Court that decides the dispute will need to ascertain the terms of the contract from the evidence given by the parties, before it can make a decision on the matters in dispute. On the other hand, if the contract terms are set out in writing in a document, which the parties subsequently sign, then both parties are bound by these terms even if they do not read them. Thus, by setting down the terms of a contract in writing, one secures the double advantage of affording evidence and avoiding disputes.

Enforcement of Contracts

An agreement can only be enforced as a contract if:

(1) the agreement relates to the future conduct of one or more of the parties to the agreement;
(2) the parties to the agreement intend that their agreement shall be enforceable at law as a contract; and
(3) it is possible to perform the contract without transgressing the law.

Validity of Contracts

The legal obligation to perform a contract exists only where the contract is valid and the following conditions are fulfilled.

(1) There must be an offer by one person (the offeror) and the acceptance of that offer by another person (the offeree), to whom the offer was made. The offer must be definite and made with the intention of entering into a binding contract. The acceptance of the offer must be absolute and be accepted in the manner prescribed or indicated by the person making the offer.
(2) The contract must have 'form' or be supported by 'consideration'. Form entails a contract made by deed and signed, sealed and delivered. Consideration is some return, pecuniary or otherwise, made by the promisee in respect of the promise made to him.
(3) Every party to a contract must be legally capable of undertaking the obligations imposed by the contract.
(4) The consent of a party to a contract must be genuine and must not be obtained by fraud, misrepresentation, duress, undue influence or mistake.
(5) The subject matter of the contract must be legal.

Remedies for Breach of Contract

Where a breach of contract occurs, a right of action exists in the Courts to remedy the matter. The principal remedies are now described.

(1) *Damages.* A breach of contract normally gives rise to a right of action for damages. The 'damages' consist of a sum of money which will, as far as practicable, place the aggrieved party in the same position as if the contract

had been performed. Where it represents a genuine estimate of the loss that is likely to be sustained, it is described as liquidated damages. If the prescribed sum is in the nature of a penalty it is unlikely to be recoverable in full.

(2) *Order for payment of a debt.* A debt is a liquidated or ascertained sum of money due from the debtor to the creditor and is recovered by an action of debt.

(3) *Specific performance.* This refers to an order of the Court directing a party to perform his part of the agreement. Its use is normally restricted to instances where damages would be an inadequate remedy, and specific performance offers a fair and practicable solution.

(4) *Injunction.* An injunction is an order of Court directing a person not to perform a specified act. For instance, if A had agreed not to carry out any further building operations on his land, for the benefit of B, who owns the adjoining land, and B subsequently observes A starting building work, then B can apply to the Court for an injunction restraining A from building. Damages, in these circumstances, would not be a suitable remedy.

(5) *Rescission.* This consists of an order of Court cancelling or setting aside a contract.

Main Characteristics of Civil Engineering Contracts

Most civil engineering contracts entered into between civil engineering contractors and their employers are of the type known as entire contracts, whereby the contractor undertakes to execute specific works for an agreed sum. In an entire contract, the contractor is not entitled to payment if he abandons the work prior to completion, and will be liable in damages for breach of contract. Where the work is discontinued at the request of the employer, or stems from circumstances which were readily foreseeable when the contract was signed and provided for in its terms, then the contractor will be entitled to be paid on a *quantum meruit* basis, that is, he will be paid as much as he has earned.

Hence the employer will usually favour the use of an entire contract to avoid the possibility of abandonment of the work prior to completion. On the other hand, most contractors will require interim payments as the work proceeds to prevent cash flow problems. For this reason the standard form of civil engineering contract[5] provides for the issue of interim certificates authorising periodic payments to the contractor.

It is customary for the contract further to provide that a prescribed proportion of the sum, due to the contractor on the issue of a certificate, shall be withheld. This sum is known as retention money and serves to insure the employer against any defects that may arise in the work. The contract remains an entire contract and the contractor is not entitled to receive payment in full until the work is satisfactorily completed, the maintenance period expired and the maintenance certificate issued.

The provision that works must be completed to the satisfaction of the employer, or his representative, does not give the employer the right to demand an unusually high standard of quality of work, in the absence of prior express agreement. Otherwise the employer might be able to postpone indefinitely his liability to pay for the works. The employer is normally only entitled to expect a standard of work that would be regarded as reasonable by competent persons with considerable experience in the class of work covered by the particular contract. The detailed requirements of the specification will have a considerable influence on these matters.

The employer or promoter of civil engineering works normally determines the conditions of contract, which define the obligations of the contractor. He often selects the contractor for the project by some form of competitive tendering and any contractor who submits a successful tender and subsequently enters into a contract is deemed in law to have voluntarily accepted the conditions of contract adopted by the employer, and any requirements embodied within the invitation to tender. For example, the employer does not usually bind himself to accept the lowest or indeed any tender. A tender is, however, normally required to be a definite offer, and acceptance of it gives rise legally to a binding contract.[2]

Types of Contract used in Civil Engineering Works

A variety of contractual arrangements are available and the engineer will often need to carefully select the form of contract which is best suited for the particular project. The employer is also entitled to know the reasoning underlying the engineer's choice of contract. The selection needs to be undertaken in a discerning and logical manner having regard to the type and size of project, the alternative approaches available with their merits and demerits, accompanied by a technical appraisal.

The principal types of contract used on civil engineering projects are now examined.

(1) *Lump Sum Contracts*

The simplest type of civil engineering contract is a lump sum contract without quantities, wherein the contractor undertakes to carry out specified works for a fixed sum of money. The nature and extent of the works are depicted on drawings, and the materials and workmanship requirements are normally detailed in a specification. No bill of quantities will be provided and so the tendering contractors often have to prepare their own quantities in order to build up their tenders.

The extent of the problems facing contractors will be influenced by the nature and degree of certainty of the work. Ideally, the use of this form of contract should be restricted to relatively small projects in which most or all of the work is above ground and clearly definable. For example, a road resurfacing contract could reasonably be on a lump sum basis, provided it did not encompass an unidentifiable quantity of making good the sub-base prior to resurfacing, otherwise the use of a schedule of rates with approximate quantities might offer a more realistic approach. The use of a lump sum contract for small projects has the advantage of saving the time and cost of preparing a bill of quantities, although problems can arise in pricing variations.[6]

This form of contract has on occasions been used where the works are uncertain in character and, by entering into a lump sum contract, the employer hoped to place the onus on the contractor for deciding the full extent of the works, and the responsibility for the payment of any additional costs, which could not be foreseen before the works were started. The employer would then pay a fixed sum for the works, regardless of their actual cost to the contractor, and this constitutes an undesirable practice from the contractor's point of view.[2] In extreme cases the lump sum may be assumed to cover all risks, including any errors in the contract documentation.

(2) *Bill of Quantities Contracts*

This type of contract, which incorporates a bill of quantities priced by the contractor, is the most commonly used form of contract for works of civil engineering construction of all but the smallest in extent, where the quantities of the work can be computed with reasonable accuracy from the drawings and associated documents. A bill of quantities is prepared in accordance with the *Civil Engineering Standard Method of Measurement* (CESMM)[7] giving, as accurately as possible, the quantities of each item of work to be executed, and the contractor enters a unit rate against each item of work. The extended totals are added together to give the tender total. This type of contract is a measure and value contract, and the contract price is the sum to be ascertained and paid in accordance with the provisions contained in the ICE Conditions of Contract.[5] Provision is made for the valuation and adjustment of rates for varied or additional work.

The preparation of comprehensive bills of quantities for civil engineering works can have an important and far-reaching effect on the cost of the works to the employer. The contractor tendering for a specific contract is provided with a schedule giving brief descriptions and quantities of all the items of work involved. In the absence of such a bill of quantities, each contractor tendering will have to assess the amount of work involved and this will normally have to be undertaken in a very short period of time in amongst other activities.

Under these circumstances a contractor, unless he is extremely short of work, is almost certain to price high to allow himself a sufficient margin to cover for any items that he may inadvertently have missed. Furthermore, there is no really satisfactory method of assessing the cost of variations and the contractor may feel obliged to make allowance for this factor also, when building up his tender.[2]

Bills of quantities assist in keeping tender figures as low as possible, as they offer the following advantages.

(1) The contractor is paid for the actual amount of work done.
(2) While providing a fair basis for payment, there is the facility for dealing with altered work.
(3) Adjudication of tenders is relatively straightforward as all tenderers price on a comparable basis.
(4) The bill of quantities gives tendering contractors a clear conception of the work involved.
(5) Most contractors in the United Kingdom are thoroughly familiar with this type of contract and are thus better able to submit a realistic price for the work.[8]

(3) Schedule Contracts

This type of contract may take one of two forms. The employer may supply a schedule of unit rates encompassing each item of work likely to be encountered and ask the contractors, when tendering, to state a percentage above or below these rates for which they would be prepared to carry out the work. Alternatively, and as is more usual, the contractors may be requested to insert prices against each item of work, and a comparison of the prices entered will enable the most favourable offer to be ascertained. Approximate quantities are sometimes included to assist contractors in pricing the schedules and the subsequent comparison of the tender figures.

This type of contract is mainly used for maintenance and similar contracts, where it is not possible to give realistic quantities of the work to be undertaken. In this form of contract it is extremely difficult to make a fair comparison between the figures submitted by the various contractors, particularly where approximate quantities are not inserted in the schedules, since there is no total figure available for comparison purposes and the unit rates may vary significantly between the different tenderers.

Another advantage of the use of schedules is that they can be prepared quickly for projects of long duration. During the execution of the early stages of a project by a contractor selected from a schedule of rates, an accurate bill of quantities can be prepared for the remainder of the work. This bill can be priced using the rates inserted in the original schedule by the contractor already employed on the site, or alternatively, competitive tenders can be obtained and, if appropriate, another contractor can carry out the later phases. For example, the substructure of a power station could be measured and valued in accordance with a schedule of rates, while the superstructure could be the subject of a bill of quantities contract. In like manner, approximate quantities could be provided for the first and possibly experimental section of a road or airport runway, with an accurate bill of quantities prepared for the main scheme.[6]

A schedule contract also enables the contractor and the engineer to co-operate at the design stage in the development of new techniques in an effective and economical way. The schedule of rates should ideally be negotiated at an early stage to give the contractor and the engineer an opportunity to discuss the relationship of plant usage and site organisation to the design of the scheme.[6]

(4) Cost Reimbursement Contracts

In cost reimbursement contracts the employer pays to the contractor the actual cost of the work plus a management fee which will include the contractor's overhead charges, supervision costs and profit. The management fee may be calculated in one of four different ways which are now described.

(i) *Prime cost plus percentage contracts.* This type of contract provides for the management fee payable to the contractor to be calculated as a percentage of the actual or allowable total cost of the civil engineering work. It permits an early starting date, as the only matter requiring agreement between the employer and the contractor is the percentage to be applied in respect of the contractor's overheads and profit. It is accordingly relatively simple to operate and was used extensively during the Second World War for defence installations, and was subject to considerable abuse on occasions.

It is a generally unsatisfactory contractual arrangement as higher costs also entail higher fees and there is accordingly no incentive for efficiency and economy. The use of this form of contract should therefore be confined to situations where the full nature and extent of the work are uncertain and urgent completion of the project is required, resulting in a critical situation. Even then every care should be taken to safeguard the employer's interests by employing a reputable contractor and arranging effective supervision of the work. The main deficiency is that an unscrupulous contractor could increase his profit by delaying the completion of the works. No incentive exists for the contractor to complete the works as quickly as possible or to try to reduce costs. Furthermore, the fee will fluctuate proportionately to any prime cost fluctuations but these will not necessarily bear any relation to any changes in the actual costs of management.[9]

A typical percentage fee might contain an addition of 100 per cent on the actual cost of wages, fares and allowances paid by the contractor to the foremen, operatives and staff (other than clerical, administrative and visiting staff) for time spent wholly on the works, together with amounts paid in respect of such wages for national insurance, graduated pensions, selective employment tax, holidays with pay, employer's liability and workmen's compensation insurance; an addition of 20 per cent on the actual cost of materials used upon the works after the deduction of all trade, cash and other discounts and rebates; an addition of 5 per cent on the actual cost of any sub-contractors' accounts in connection with the works and any payments made by the employer; and an addition of 10 per cent on the actual cost of any mechanical plant used on the site upon the works.[9]

(ii) *Prime cost plus fixed fee contracts.* In this form of contract the sum paid to the contractor will be the actual cost incurred in the execution of the works plus a fixed lump sum, which has previously been agreed upon and does not fluctuate with the final cost of the project. No real incentive exists for the contractor to secure efficient working arrangements on the site, although it is to his advantage to earn the fixed fee as quickly as possible and so release his resources for other work. This type of contract has advantages over the prime cost plus percentage contract from the employer's standpoint.[2]

In order to establish a realistic figure for the fixed fee, it is necessary to be able to assess with reasonable accuracy the likely amount of the prime cost at the tender stage, otherwise the employer may have to revert to a prime cost plus percentage contract with its inherent disadvantages. It is advisable to prepare a document showing the estimated cost of the project in as much detail as possible so that the work is clearly defined and also the basis on which the fixed fee is calculated.

(iii) *Prime cost plus fluctuating fee contracts.* In this form of contract the contractor is paid the actual cost of the work plus a fee, with the amount of the fee being determined by reference to the allowable cost by some form of sliding scale. Thus, the lower the final cost of the works (prime cost), the greater will be the value of the fee that the contractor receives. An incentive then exists for the contractor to carry out the work as quickly and cheaply as possible, and it does constitute the most efficient of the three types of prime cost contract that have so far been described.[2]

(iv) *Target cost contracts.* These are used on occasions to encourage the contractor to execute the work as cheaply as possible. A basic fee is generally quoted as a percentage of an agreed target estimate usually obtained from a priced bill of quantities. The target estimate may be adjusted for variations in quantity and design, and fluctuations in the cost of labour and materials and related matters. The actual fee paid to the contractor is determined by increasing or reducing the basic fee by an agreed percentage of the saving or excess between the actual cost and the adjusted target estimate. In some cases a bonus or penalty based on the completion time may also be applied.[2]

Hence prime costs are recorded and a fee agreed for management services provided by the contractor as in the other forms of cost reimbursement contract. The actual amount paid to the contractor depends on the difference between the target price and the actual prime cost. In practice, various methods have been used for computing this sum. An alternative method that has been used is to pay the contractor the prime cost plus the agreed fee, and for the difference between target price and prime cost, whether a saving or an extra, to be shared between the employer and the contractor in agreed proportions. Yet another method is to pay either the target price or the prime cost plus the agreed fee whichever is the lower. This latter form of contract does, in fact, combine the characteristics of both the fixed price and cost reimbursement contracts.[9]

Fluctuations in fee due to differences between target and actual costs operate as a bonus to the contractor if his management is efficient, or as a penalty if it is inefficient. The benefits to be obtained by the employer from this contractual arrangement are mainly dependent on the target price being agreed at a realistic value, as there will be a great incentive for the contractor to increase the estimated price as much as possible in the first instance. It is essential that the employer obtains expert advice in evaluating this price. It may be negotiated with the contractor or established in competition. Target cost contracts should not be entered into lightly as they are expensive to manage, and require accurate management and careful costing on the employer's behalf.[10]

(5) All-in Contracts

With this type of contract the employer or promoter, frequently using the services of an engineer, normally gives his requirements in broad outline to contractors, who are asked to submit full details of design, construction and cost, and probably including maintenance of the works for a limited period. This procedure has been used in the chemical and oil industries and for the design and construction of nuclear power stations for the Central Electricity Generating Board. It is a contractual arrangement which sometimes finds favour with overseas employers. Although they may appear very attractive, they may give rise to many difficulties in implementation. As an ICE report[4] emphasises, they can prove particularly difficult to operate in times of high inflation and when the basic technology is changing rapidly.

All-in contracts are sometimes referred to as package deals and, in practice, the arrangements may vary considerably, ranging from projects where the contractor uses his own professional design staff and undertakes both complete design and construction, to projects where the contractor, specialising in a particular form of construction, offers to provide a full service based on preliminary sketch plans provided by the employer's engineer. All-in contracts can be on a fixed price or cost reimbursement basis, competitive or negotiated, and can incorporate the management contracting system described later in the chapter. The employer may require the contractor to finance the project until it is revenue producing, in which case it is often referred to as a turnkey contract.[10]

The selection of the contractor should be based on a brief of the employer's requirements. The brief should ideally be prepared by the employer's engineer and costed by him, so that contractors are tendering on a brief that is within the employer's budget. It is costly for contractors to tender for this type of contract in competition, as each contractor will have to produce a design to meet the brief and price for construction. Where this process is taken to excess at the tendering stage, it will result in an uneconomic use of resources. Hence, many contractors are not prepared to proceed beyond outline sketch design and an indicative price at the competitive tender stage. Furthermore, the evaluation and comparison of contractors' tenders is complicated as each contractor is likely to interpret the brief in a different way. Hence considerable adjustments may be needed to reduce them to a common basis for purposes of comparison.[10]

(6) Negotiated Contracts

As a general rule, negotiation of a contract with a single contractor should take place only if it can be shown to result in positive advantages to the employer. There are a number of situations in which negotiations may be beneficial to the employer and some of the more common instances are now listed:

(1) The employer has a business relationship with the contractor.
(2) The employer finds it difficult, or even impossible, to finance the project in any other way.
(3) The employer has let a contract in competition, and then another contract of similar design comes on programme.
(4) In particular geographical areas where there may be only one contractor available to do the work.
(5) A certain contractor is the only one available with either the expertise or the special plant required to carry out the project.
(6) At times when the construction industry is grossly over-stretched and negotiation offers the best approach.
(7) Where a rapid start is required, as for example when the original contractor has gone into liquidation.[10]

The two principal methods of negotiation are:

(1) Using the competitive rates obtained for similar work undertaken under similar conditions in another contract; but there are many inherent problems in adjusting the existing rates to provide a basis for pricing the new work.
(2) An agreed assessment of the estimated cost, to which will be added an agreed percentage for head office overheads and profit, which can be subsequently documented in a normal bill of quantities.

There are certain essential features which are required if the negotiation is to proceed satisfactorily. These include equality of the negotiator for each party, parity of information, agreement as to the basis of negotiation and an approximate apportionment of cost between suitable heads, such as site management, contractor's own labour, direct materials, plant, contractor's own sub-contractors, nominated sub-contractors, nominated suppliers, provisional sums and contingencies, and head office overheads and profit.[10]

Advantages of negotiation. Advantages can accrue from a decision to select a contractor and to negotiate a contract sum with him. For instance, the contractor can be brought in at an early stage as a member of the design team, so that full advantage can be taken of his knowledge, experience and constructional resources. He can take an effective role in the planning process, which should help towards producing a better design solution at lower cost, and possibly with a shorter completion time. Further benefits may be secured if sub-contractors are brought in at the same time.

With the contractor appointed, agreement may be reached on the format of the bill of quantities which will be of the greatest use to the contractor in programming, progressing and cost controlling the project. Where bills are produced by computer, it is relatively easy to produce bills in any alternative form very quickly.[9]

Contractor selection. The selection of the contractor can be carried out in a number of different ways and the following methods are used extensively:

(1) appointment of a contractor who the employer and his professional advisor(s) believe can carry out the work to a satisfactory standard and a fair price;

(2) appointment of a contractor already employed on a similar project; pricing the bill of quantities at the same rates as in the existing contract with suitable adjustments;

(3) selection of a contractor following the interview of a number of suitable contractors, agreement having been reached on the percentage to be added to the estimated nett cost for overheads and profit; and

(4) selection of a contractor following the submission in competition by a number of contractors, whose ability to execute the contract efficiently has been established, of a document which shows in considerable detail the method to be employed in pricing the bill of quantities when it becomes available.[9]

Procedural aspects. Contractors' estimating methods vary widely, as shown in chapter 3, and so the tender documents should be presented in a way that permits maximum flexibility. The information normally falls under these heads:

(1) Conditions of Contract.

(2) Information relating to the build-up of effective labour rates, productivity factors and examples demonstrating how these should be applied in calculating unit rates. All the major items of work which the contractor is expected to undertake will be included, with the approximate quantity of each stated.

(3) Information on the pricing of insurances, attendances and the like, and the percentage each contractor will require to be added to the estimated nett prices for overheads and profit.

There are, however, weaknesses inherent in the negotiating process, such as the length of time required for pricing and negotiation and that there is rarely any guarantee that a lower price will be obtained than by the normal competitive tendering procedure. It may be argued that the allowance for business risk is a matter of opinion, the anticipated profit is based on hope, and off-site overheads are dictated by the efficiency of the construction organisation. Hence the probability of negotiating a contractor's margin equal to, or less than, that prevailing in the competitive market is in all probability unlikely.[9]

(7) *Management Contracts*

The management contract is a system whereby a main contractor is appointed, either by negotiation or in competition, and works closely with the employer's professional advisor(s). All physical construction is undertaken by sub-contractors selected in competition. The management contractor provides common services to the sub-contractors such as welfare facilities, and plant and equipment that is not confined to one sub-contractor, and sufficient management both on and off the site to undertake the planning and management, co-ordination and control of the project. He is paid a fee for his services and in addition, the cost of his on-site management, common services and the cost of all work undertaken by sub-contractors.[11]

Uses. The management contract, which emanated from the United States, is most appropriate to large, complex projects exceeding £20m in value, which exhibit particular problems that militate against the employment of fixed price contract procedures. Typical examples are:

(1) Projects for which complicated machinery is to be installed concurrently with the construction work.
(2) Projects for which the design process will of necessity continue throughout most of the construction period.
(3) Projects on which construction problems are such that it is necessary or desirable that the design and management team includes a suitably experienced contractor appointed on such a basis that his interests are largely synonymous with those of the employer's professional advisor(s).[12]

Procedures. A wide range of views exists as to the best procedures to be adopted in management contracting, but they usually incorporate the following activities and requirements.

(1) The management contractor is precluded from carrying out any of the physical works using directly employed labour. His role is primarily that of planner, manager and organiser.

(2) The works are divided into packages agreed by the employer's professional advisor(s) and the management contractor as being the most appropriate for the particular project. Competitive tenders are invited for each package from tenderers selected by the employer's professional advisor(s).

(3) The management contractor provides from his own resources:
  (a) Site supervisory, technical and administrative staff to run the contract.
  (b) Those facilities to be shared by the sub-contractors where they are not included in any of the agreed sub-contract packages.

(4) The management contractor is normally paid by monthly instalments computed on the following basis:
  (a) The substantiated amounts due to be paid by him to sub-contractors.
  (b) The nett cost to him of providing site staff and shared facilities.
  (c) A management fee which may be in two parts:
    (i) a pre-commencement management fee, which will be a lump sum; and
    (ii) a construction management fee which is a percentage of the nett costs to the management contractor, including payments to be made by him to sub-contractors. The percentage is usually in the order of 4 to 5 per cent.

The management contractor will be concerned with keeping the cost of the works within the project budget, reporting to the employer on possible extras and also in dealing with the sub-contractors in regard to such matters as claims for loss and expense and the settlement of accounts. The management contractor attends all design and progress meetings and it is good policy for a representative of the employer also to be in attendance. The management contractor will be able to report, among other matters, on the dates by which he will require design information and on any information that is already late.[12]

PRELIMINARY INVESTIGATIONS

Preliminary investigations relating to civil engineering projects can take a variety of forms. An ICE report[4] describes how investigations can encompass non-technical matters, such as economic aspects, and demographic or social factors, or involve purely technical investigations of which the most common incorporate site exploration work.

Site Exploration Work

A thorough site investigation is an essential requisite for the efficient planning and execution of any civil engineering project. Existing geological maps and data normally provide the first source of reference. In addition to any special investigations, such as resistivity and seismographic surveys, soil surveys are often undertaken by specialists who provide borehole logs, samples, laboratory test results and a report containing an interpretation of the results by a person experienced in geotechnical matters. Nevertheless, the engineer for the project makes the final assessment of the results in so far as they affect the design and specification of the works.

In addition, there may be other matters requiring investigation, such as the nature and condition of access to the site, any statutory restrictions which can affect the design and/or construction of the works, the rights of adjacent land-owners such as rights of light and way and pipe easements, the availability and capacity of sewers, water, gas and electricity services, availability of materials required for construction, ease of disposal of surplus materials, groundwater level, and liability to flooding and subsidence.

General Procedure

Cottington and Akenhead[13] have described how decisions in the Court of Appeal (*Batty v. Metropolitan Property Realisations Ltd 1978* and *Acrecrest Ltd v. W.S. Hattrell and Partners 1982*) show clearly that the developer, who is usually the employer or promoter, owes a duty in law, independent of contract, to examine the land on which he intends to build to ensure that it is suitable for the intended purpose.

Thus the employer would be taking an unacceptable risk if he failed to carry out an adequate ground investigation. The employer is also responsible for ensuring that all statutory consents, such as planning and building regulations approvals are obtained prior to letting the contract. The *ICE Conditions*[5] render the employer liable to the contractor for acts or omissions of the engineer which are not reasonably foreseeable and which cause the contractor to incur delays and additional costs.[13]

The engineer owes to the employer duties in contract and in tort to exercise all reasonable care and skill in the provision of civil engineering services, including the preliminary investigations. He must not claim to have expertise which he does not possess. In this context, the engineer may recommend to the employer that a contractor be appointed to carry out a full site or ground investigation prior to the formulation of the detailed design of the project.

In a ground investigation, the engineer or his representative will supervise, monitor and, if necessary, vary the form of investigations in the light of the initial findings. He must have sufficient experience to recognise critical changes in soil type or conditions which will not only affect the nature of the investigation but also the design of the permanent works.[13]

It will be necessary for the engineer to define clearly the information to be provided by the investigation. For example, it may be that:

(1) test results only are required; or
(2) intermediate and final reports are needed; or

(3) reports are required to contain specific recommendations with regard to land use, the type of drainage, roads and foundations that would be most appropriate, and an indication of any specific hazards which are likely to be encountered.[13]

Potential hazards could include restrictions relating to safety during and after construction, embracing support to excavations, the effect on adjacent properties and the effect of contaminants, including flammable, toxic and radioactive substances. Tenderers must have sufficient qualified and experienced geotechnical staff, laboratories and site operatives.[13]

Cottington and Akenhead[13] have described how where the employer has little or no geotechnical experience, it would be advisable for the engineer to give reasons for the ground investigation possibly in narrative form. The following example will serve to illustrate this approach.

Proposed Development at Snowley

The proposed development consists of light industrial and residential development on the eastern outskirts of Snowley. The development site comprises plots (parcels) 5914, 5915, 5916 and 5919 as shown on Ordnance Survey map, sheet nr 124, with a total area of 153 acres (62 hectares). As far as can be established, the area to be developed has always been pasture with the exception of plot 5915, which has been under continuous cultivation for at least 60 years.

Subject to the findings of a ground investigation, it is proposed to develop plots 5914 and 5915 with light industry and plots 5916 and 5919 with residential properties. It is envisaged that the residential development will consist of three-storey flats, two-storey semi-detached houses and groups of single storey accommodation for elderly persons at an approximate overall density of 12 residential units to the acre (29.6 to the hectare).

The investigation is required to determine the engineering properties of the soil and the groundwater characteristics on which to base the design and construction of subsoil, surface and foul drainage, road construction to withstand single wheel loads of up to 5 tonnes and foundation design for all the constituent structures. Information is also required on all hazards to health and safety within the site both during and on completion of construction.

It is proposed to instal foul and surface water pumping stations on plot 5916 with rising mains in the approximate locations shown on the Ordnance Survey map.

The details of the internal road layout have yet to be determined, but access to the site will be from the B4635 through plot 5915.

Contractual Requirements for Ground Investigations

The principal contractual requirements relating to this specialised form of contract are contained in the *ICE Conditions of Contract for Ground Investigation.*[14]

A summary of the principal general obligations of the contractor undertaking a ground investigation contract follows.

General Responsibilities

(1) The contractor shall, subject to the provisions of the contract, carry out the investigation and provide all labour, materials, equipment, instrumentation, transport to and from and in or about the site, and everything, whether of a temporary or permanent nature, required in and for the investigation so far as the necessity for providing the same is specified in, or reasonably to be inferred from, the contract.

(2) The contractor shall take full responsibility for the adequacy, stability and safety of all site operations, laboratory testing and methods of working.

(3) The contractor shall submit to the engineer the name and address of the laboratories undertaking testing in accordance with the contract and shall obtain the engineer's approval in writing before the services and equipment of such laboratories may be used in the execution of the contract.

(4) Except as may be expressly provided in the contract, the contractor shall not be responsible for the design or specification of any ancillary works required to be installed or constructed in accordance with the contract.

Inspection of Site

The Contractor shall be deemed to have inspected and examined the site and its surroundings and to have informed himself before submitting his tender as to the general nature of the geology (so far as is practicable and having taken into account any information in connection therewith which may have been provided by or on behalf of the employer), the form and nature of the site, the extent and nature of the work and materials necessary for the completion of the investigation, the means of communication with and access to the site, the accommodation he may require, and in general to have obtained for himself all necessary information as to risks and contingencies and all other circumstances influencing or affecting his tender.

Programme to be Furnished

Within 21 days after the acceptance of his tender, the contractor shall submit to the engineer for his approval a programme showing the order in which he proposes to carry out the investigation and thereafter shall furnish such details and information as the engineer may reasonably require in regard thereto. The contractor shall at the same time also provide in writing for the information of the engineer a description of the arrangements and methods which he proposes to adopt for carrying out the investigation.

Care of Samples and Cores

The Contractor shall, unless it is otherwise provided for in the contract, take full responsibility for the care and storage of the samples and cores obtained from the investigation at his cost until 28 days after the issue of the report to the engineer or in the case of a phased investigation the relevant section of the report. After the said period the contractor shall give the engineer 14 days written notice of his intention to charge rental for the storage of the cores and samples. The engineer shall then either give instructions for the immediate disposal of the samples and cores at the contractor's expense or state his storage or other requirements giving an indication of his programme.

Other Contractual Aspects

This form of contract also contains many of the customary general requirements, covering such matters as the powers of the engineer's respresentative, restrictions on assignment and sub-letting, supply of contract documents, provision of sureties, additional cost incurred through adverse physical conditions and artificial obstructions, delay and extra cost, setting out, safety and security, insurance, reinstatement, damage to persons and property, statutory requirements, quality of workmanship and materials, variations, completion certificate, maintenance period, property in materials and plant, measurement and valuation, certificates and payment and settlement of disputes.

Nature and Scope of Ground Investigations

These matters are well described and illustrated in an ICE Works Construction Guide,[15] to which readers are referred for more detailed information and further references. Site investigations for civil engineering projects can range from a walk over the site and examination of records, probably supplemented by the excavation of some trial pits, to a comprehensive programme of borings, testing and analysis, normally involving the services of both civil engineers and geologists.

Preliminary Investigations

The investigation usually starts with a desk study of all relevant Ordnance Survey maps and geological maps and records, such as those obtainable from the Institute of Geological Sciences and the National Coal Board. Valuable information may also be available from past work on the site or on adjacent properties. This study should indicate the general characteristics of the site, such as the superficial and solid geological formations and associated soil and rock types, the likelihood of structural discontinuities and sub-surface topographical variations.

    A visual inspection of the site will identify the main topographical and geological features, such as the presence of man-made fill deposits, peat or other compressible soils or conversely of shallow rock, and the existence of a high water table.[15]

Pits, Borings and Drillings

(a) *Pits.* The simplest and often the most satisfactory method of sub-surface investigation is by excavating trial pits, either by machine or manually in restricted locations, as they permit a visual inspection of soil and ground-water conditions. It is, however, necessary to take some basic precautions.

   (1) They result in ground disturbance and should be sited away from areas where foundations may be constructed.
   (2) They give no indication of conditions below the level of excavation, which must be assessed by geological inference or by boring.
   (3) The sides of the pits must be adequately supported for the safety of personnel.[15]

(b) *Borings.* A variety of boring techniques are used in site investigation work, but in soft formations the 'cable and tool' or 'shell and auger' are the most commonly used forms of equipment. Steel casings are often driven to line the borehole as it proceeds downwards. A nominal minimum diameter of 150 mm is used to permit 100 mm diameter samples to be obtained.

   The normal boring tools consist of an open ended tube or 'clay cutter', and a 'shell' or 'clack-pump', which has a flap to retain sand and gravel immersed in water. Chisels are used to break up coarse material prior to its removal by the shell. The boring equipment operates from a winch and tripod or a four-legged pulley frame.

(c) *Rotary drilling.* Power augers may be used for rapid penetration of cohesive soils and weak rock, but are rarely suitable for gravel, sand or silt below water level, or for soils containing boulders. Drilling of rock may be carried out with skid-, lorry- or track-mounted rotary drills.

Sampling

Representative samples are required from every stratum penetrated in trial pits and boreholes, and at regular intervals from a seemingly uniform stratum. The sample containers should be labelled immediately with details of the site, date, pit or borehole number and depth. With certain laboratory tests it is necessary to take large bulk samples of 25 kg or more.

The main source of material for laboratory tests of cohesive soil is undisturbed samples. However, in practice, some disturbance is almost certain to occur, the extent of the disturbance being influenced by the soil type, design of sampler and the care and method used by the operator.

Groundwater

The supply of groundwater data forms an important aspect in any ground investigation, and normally covers both head and flow. The piezometric head or standing water level is the level to which the water will eventually rise and may be influenced by the season and adjoining tidal conditions. The flow of water into a borehole or pit below water level is affected by the dimensions of the excavation and the permeability of the soil. All water strikes are recorded and also water levels at the beginning and end of each shift.

Laboratory Testing

A wide range of mechanical and chemical tests are frequently used in foundation engineering work. Mechanical tests are concerned primarily with strength and consolidation, while chemical tests are mainly aimed at determining possible aggression to Portland cement products (pH and sulphate content), and are usually carried out on groundwater samples.

Procedural Aspects

A flexible approach is recommended by Robb,[15] with an individual approach to each site. A pattern of trial pits or boreholes will be planned having regard to the probable location of the proposed structures. In the absence of this information a grid layout is likely to be adopted, often with a spacing of about 20 to 25 m. There should be at least three boreholes on even the smallest sites to provide representative data.

The depth of boring is determined by the anticipated depth of impact of the foundations, and there should ideally be at least one relatively deep hole on each site. Where piled or deep foundations are envisaged, investigations should proceed to some 5 m below the probable toe level.

Norgrove and Attewell[16] have described how a fundamental contribution to the effectiveness of a ground investigation for tunnelling is provided by the spatial density of boreholes, as with long linear works the scope for geological and hydrological change increases. A borehole should be located at the centre of each access shaft/manhole position, with additional holes being offset by approximately one tunnel diameter from the walls of and at either side of the tunnel line. A representative figure for the cost of site investigation for tunnelling was found to be between 0.75 and 1.5 per cent of the total project cost and would not normally exceed 2 per cent.[16]

For tunnelling in soil, the two most important objectives at the ground investigation drilling and logging stage were identified as:

(1) careful definition of the composition and fabric of the soil, particularly within and just above the tunnel section; and
(2) detection of free groundwater and definition of rest water levels.[16]

Reports and their Interpretation

A ground investigation report normally starts with a statement of the reasons for the investigation and the method of carrying it out. The results of the desk study are summarised in addition to those of the ground investigation, often supported by relevant graphs and tables, leading into the broad conclusions on which design and construction decisions will be based. Records of boreholes and test results are often incorporated as appendices to the report. A typical record of four boreholes is illustrated in figure 1.1, and these often form the most important part of a ground investigation report. Furthermore, each word of a soil or rock description, except those relating to colour, can be interpreted against appropriate British Standard Codes of Practice, such as BS 5930.[17]

The report will be of considerable value to a number of persons. For instance, the foundation designer will need to know the strength and compressibility for bearing capacity and settlement calculations, together with groundwater levels for consideration of flotation. The earthworks contractor will be concerned with the relative difficulty of excavation, suitability of materials for re-use and effect on groundwater of excavating below the water table. From the report, the piling contractor will determine the most economic system, the level to which driven piles will penetrate or the depth of drilled shaft which must be supported by casing, while the drainage contractor can decide the depth and spacing of wells or wellpoints from particle size, permeability and water level data.[15]

ENGINEERS' REPORTS

An ICE Report[4] has described how the format, length and detailed content of an engineer's report may vary according to the size and complexity of the project, the number of alternative solutions to be examined, and the nature and scope of the decision-making procedures involved before the scheme can be implemented. For some projects, a single report describing the recommended design and including its estimated cost, and possibly also the cost of operation after completion, will be sufficient. Other schemes may require a series of separate reports, each possibly comprising a number of volumes. Some reports on civil engineering projects are prepared wholly by engineers, while others may be formulated by a team of specialists of disparate disciplines. The same ICE publication[4] describes how a report detailing the preferred solution with a phased construction programme for a major project may comprise some or all of the following.

(1) *Sector study* — assesses the requirements of a particular area or field and aims to identify individual projects for investigation.
(2) *Pre-feasibility study* — investigates the viability of demand, whether the required resources are likely to be within acceptable cost limits, and whether or not a feasibility study is justified.

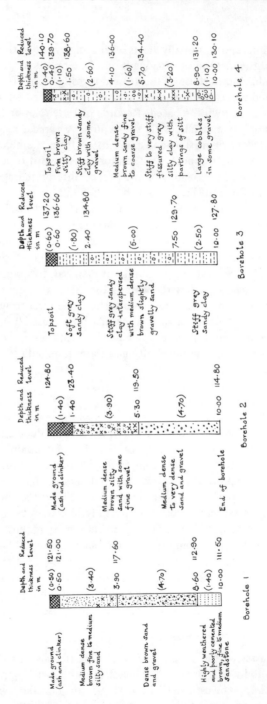

Figure 1.1 Records of boreholes on sewage treatment plant site

(3) *Feasibility or pre-investment study* – comprises preliminary surveys to investigate technical and economic feasibility, an estimate of capital and operating costs, and such additional information as is required to enable the employer to decide whether or not he should endeavour to finance the project. It may include some outline design proposals.

(4) *Master plan* – a long-term development programme showing how construction and expenditure can be phased.

(5) *Environmental impact statement* – examines the effect of the proposed development on the environment.

(6) *Project report* – develops the recommended solution in more detail, with particular reference to its technical aspects. It is sometimes referred to as the final design report.

(7) *Geotechnical report* – as described earlier in the chapter, this investigates the nature of the ground on which the works are to be constructed and assesses possible problems.

(8) *Financing report* – where the employer requires the engineer to identify possible sources of finance, to assist him in making arrangements for the provision and repayment of the necessary funds.

Objectives

The engineer following his investigations will submit a report to the employer describing concisely his investigations and the main conclusions. He will often outline the scheme which he believes will best meet the employer's requirements, together with particulars of the financial implications, to assist the employer in making a realistic decision.

Presentation

The format and presentation of the engineer's report will vary according to the type of project and the purpose for which it was commissioned. However, it generally provides a review of the investigations undertaken by the engineer, compares the possible alternative solutions on technical, economic and financial grounds, and then proceeds to make appropriate recommendations. The report must be written in straightforward, clear, concise and unambiguous terms so that it can be readily understood by the employer and any others who may refer to it, as they may not be able to appreciate the finer technical detail nor have adequate time to assimilate it. Nevertheless, the report must contain sufficient technical information to convince any other engineer or specialist of the facts and of the soundness of the engineer's judgement.[4]

Format

The report should set out in a clear and concise manner the engineer's brief, the action he has taken and his conclusions. The most effective approach is often to summarise the salient points in the first section and to incorporate the technical details in appendices or in separate volumes. The ICE publication *Civil Engineering Procedure*[4] describes how in most cases the main body of the report falls into two parts.

The first part contains:

(1) title page
(2) letter of transmittal from the engineer to the employer
(3) table of contents
(4) introduction, which often includes a summary of the scope of the work, statement or terms of reference
(5) summary of investigations
(6) summary of conclusions and recommendations.

The second part, comprising the main body of the report, contains:

(1) terms of reference
(2) statement of the problem, its significance, scope and history
(3) basic assumptions, data or trends on which the engineer's investigations have been based, and his reasons for adopting them
(4) details of investigations carried out
(5) design and other criteria used
(6) alternative solutions considered
(7) cost estimates
(8) comparison of options considered, and reasons for rejection of non-preferred schemes
(9) description of the recommended solution, programme for implementation, conclusions and recommendations.

Practical Applications

A report of a highway engineer might, for example, be concerned with the comparison of the costs of improving an existing road adhering to its original alignment with an alternative proposal whereby the road is realigned and reconstructed. For instance, assuming the existing road is 40 km in length and is estimated to cost £1.2m in improvement works, while the reconstructed realigned road will be 35 km long and is estimated to cost £1.7m.

Although the reconstructed realigned road is considerably more expensive in first cost than the improved existing road, the length is reduced by 12.5 per cent with consequently lower annual maintenance costs. To determine the most economic long-term solution, it will be necessary to compare the discounted maintenance costs for a 20-year period, and to assess the benefits accruing to users of the improved and more direct road. It is possible that these savings and benefits will more than offset the additional capital cost of the realigned road. Admittedly, this example introduces the difficult and imprecise technique of cost benefit analysis which must be applied with care.

An engineer reporting on the reconstruction and extension of a sewage purification works will describe the deficiencies of the existing works with regard to construction, capacity, operation and quality of effluent. He will endeavour to forecast future population trends and flows through the works and any likely changes in the type of sewage to be treated as a result of new industrial developments or for other reasons.

He will then proceed to describe how parts of the existing works can be renovated and retained, possibly even using them for a different purpose. The new structures and plant will be described showing clearly their principal advantages in more effective operation with superior standby facilities and the provision of reserve capacity.

The new installations will be shown to produce savings in operating costs as a result of reduced manual operations. The upward flow sedimentation and humus tanks will be more efficient and also occupy less space on a restricted site. A system of secondary digestion of sludge will overcome the problems that currently arise with an excess of partly dried sludge in open sludge drying beds.

The engineer will provide an estimate of the cost of reconstruction and extension and indicate the proposed loan repayment period, which is usually in the order of 60 years. The estimated cost will be broken down into its component parts, with additional sums included to cover contingencies, engineer's fees and printing, resident engineer's salary and any land, easements, legal and associated costs.

REFERENCES

1. Institution of Civil Engineers. *Royal Charter, By-laws, Regulations and Rules* (1982)
2. I.H. Seeley. *Civil Engineering Quantities.* Macmillan (1977)
3. Institution of Civil Engineers. *Present Organisation* (1982)
4. Institution of Civil Engineers. *Civil Engineering Procedure* (1979)
5. Institution of Civil Engineers, Association of Consulting Engineers and Federation of Civil Engineering Contractors. *General Conditions of Contract and Forms of Tender, Agreement and Bond for use in connection with Works of Civil Engineering Construction.* Fifth Edition (June 1973, revised January 1979)

6. R.J. Marks, R.J.E. Marks and R.E. Jackson. *Aspects of Civil Engineering Contract Procedure.* Pergamon (1985)
7. Institution of Civil Engineers and Federation of Civil Engineering Contractors. *Civil Engineering Standard Method of Measurement* (1985)
8. C.K. Haswell and D.S. de Silva. *Civil Engineering Contracts.* Butterworths (1982)
9. I.H. Seeley. *Quantity Surveying Practice.* Macmillan (1984)
10. The Aqua Group. *Tenders and Contracts for Building.* Granada (1982)
11. H. Davis. The advantages of management contracts — apparent or real? *Proceedings of the Royal Institution of Chartered Surveyors, Quantity Surveyors Twelfth Triennial Conference* (April 1981)
12. Q.S. Digest. Management contracts. *The Surveying Technician* (June 1982)
13. J. Cottington and R. Akenhead. *Site Investigation and the Law.* Telford (1984)
14. Institution of Civil Engineers, Association of Consulting Engineers and Federation of Civil Engineering Contractors. *ICE Conditions of Contract for Ground Investigation.* Telford (October 1983)
15. A.D. Robb. *Site Investigation.* Telford (1982)
16. W.B. Norgrove and P.B. Attewell. Assessing the benefits of site investigation for tunnelling. *Municipal Engineer* (August 1984)
17. British Standards Institution. *Code of Practice for Site Investigations: BS 5930* (1981)

2 Contract Documentation

This chapter examines the form and purpose of the various contract documents and the detailed requirements and implications of the ICE Conditions of Contract.

FORM AND PURPOSE OF CONTRACT DOCUMENTS

Contract documents form the basis on which a civil engineering contractor will prepare his tender and carry out and complete the contract works. It is, accordingly, essential that the documents shall collectively detail all the requirements of the project in a comprehensive and unambiguous way. These documents also identify all the rights and duties of the main parties to the contract – the employer, engineer and contractor. Collectively they constitute a binding contract, whereby the contractor undertakes to construct works in accordance with the details supplied by the engineer and the employer agrees to pay the contractor in stages during the execution of the works in the manner prescribed in the contract.

The documents normally used in connection with a measurement contract are listed in *Civil Engineering Procedure*,[1] and consist of the following items, which are all mutually operative.

(1) The Conditions of Contract

These commonly referred to as the ICE Conditions,[2] define the terms under which the work is to be undertaken, the relationship between the employer and the contractor, the powers of the engineer and the terms of payment. These Conditions have been prepared jointly by the Institution of Civil Engineers, the Association of Consulting Engineers and the Federation of Civil Engineering Contractors, and are designed to protect the interests of all the parties involved.

The use of these standard conditions ensures that all the parties are familiar with their terms as a result of common usage and there is less likelihood of disputes arising over the interpretation of contract clauses. The employer is fully aware of the arrangements for payment to the contractor on certification by the engineer. The engineer has extensive powers of control and direction of the works as prescribed in the Conditions, but he must exercise his duties reasonably. The duties and obligations of the contractor are well documented and these will be considered in more detail later in the chapter.

It is unwise to make modifications to the ICE Conditions as these can lead to uncertainties and disputes which are better avoided. They also make the task of the contractor in preparing his tender more difficult as he faces unfamiliar conditions, and has to assess their financial implications in the restricted time available for tendering. Some public authorities have introduced modifications to the ICE Conditions to meet their own particular requirements. An ICE publication[1] emphasises the desirability of adding special conditions, where necessary, instead of making modifications to the standard conditions. In any event, all amendments and additions must be clearly drawn to the notice of tendering contractors.

The standard ICE Conditions are examined in detail later in this chapter.

(2) Drawings

The drawings are prepared by the engineer to clearly depict the nature and scope of the contract works. They should be as comprehensive and detailed as possible to assist the contractor in compiling his tender. On occasions it may not be practicable to supply full details at the tendering stage, but even then as much information as possible should be given, and there must certainly be adequate information to enable the tendering contractors to understand what is required.

All available information should be provided relating to site conditions and in particular data regarding soils and groundwater, as described in chapter 1. Drawings may incorporate or be accompanied by schedules, such as those recording details of steel reinforcement and manholes.

For example, a contract encompassing a sewage treatment works, sewers and other associated work could be depicted on the following drawings: layout of sewage treatment works; working drawings of each section of the works, such as siteworks, inlet works, primary tanks, aeration tanks, secondary tanks, pumphouses, and culverts and pipework to a scale of 1:50 or 1:100; layout of the sewers and manholes often on a 1:2000 plan; longitudinal sections of sewers often to an exaggerated combined scale of 1:2000 horizontally and 1:500 vertically; manhole details to a scale of 1:50; and pumphouse details, where appropriate, to 1:50 scale.

All drawings should contain ample descriptive and explanatory notes which should be legible and free from abbreviations, and be suitably referenced. Ample figured dimensions should be inserted on the drawings to ensure maximum accuracy in taking off quantities and in setting out constructional work on site.

Existing and proposed work must be clearly distinguished on the drawings. For instance, old and new sewers and other services can be depicted in different colours or different types of line. With alterations to buildings it is often desirable to prepare separate plans of old and new work to avoid any confusion.

Materials shown in section are best hatched for ease of identification, using the notations shown in BS 308 (Engineering Drawing Practice) and BS 1192 (Construction Drawing Practice). Guidance is also given in the British Standards on the use of lines for various purposes: for instance, dimension lines are to be thin and continuous with the dimensions placed above the line and along it, and reading from the bottom or right hand edge of the drawing. It is good practice to keep a card index of drawings and to enter on drawings the date and nature of any amendments.

During the progress of the works, the engineer or his representative often finds it necessary to issue further drawings in amplification of those issued at the tender stage. Some may give details of components which were the subject of prime cost or provisional sums in the bill of quantities.

(3) Specification

The specification is another important document in a civil engineering contract. It describes in detail the work to be executed, the character and quality of the materials and workmanship, and any special responsibilities of the contractor that are not covered by the Conditions of Contract.[2] It is often subdivided into two parts as illustrated in *Civil Engineering Specification*[3] — one dealing with general information and the general duties and obligations of the contractor not specifically covered by the Conditions of Contract, and the other specifying materials and workmanship requirements, and normally entered in the sequence of work sections as listed in the *Civil Engineering Standard Method of Measurement.*[4] Alternatively, the materials and workmanship requirements may each be entered separately for the entire project.

A good arrangement for a specification covering civil engineering works is to start with any special conditions relating to the contract and the extent of the contract; then to follow with a list of contract drawings, description of access to the site, supply of electricity and water, provision of offices and mess facilities and statements regarding suspension of work during frost and bad weather, damage to existing services, details of borings, water levels and similar clauses.

The specification often lays down the order in which the various parts of the work are to be executed, the methods to be adopted, and particulars of any facilities which are to be afforded to other contractors. It generally requires the contractor to submit a programme and details of associated temporary works. Care is needed when drafting a specification to avoid any conflict with provisions in the Conditions of Contract or bill of quantities. Furthermore, all the specification clauses should be clear, precise, complete and up-to-date, suitably numbered for ease of reference and with adequate headings and sub-headings to act as signposts to the contractor.

Constant references to British Standards and Codes of Practice are usually made in the specification, and they must refer to the latest publications and, where appropriate, the relevant classes or grades of materials or components. The engineer should not specify a quality which is clearly far in excess of the standard required for the particular project.

Table 2.1 contains typical specification clauses relating to tunnel and shaft linings to illustrate the usual format and form of wording.

Table 2.1 Specification of tunnel and shaft linings

1. *Shaft segments*
   Shaft rings shall be of cast iron of 7 m internal diameter, with each ring 450 mm long and consisting of ten ordinary plates, two top plates and one key. The approximate weight of each ring is 4.4 tonnes. Each plate shall be provided with a 30 mm diameter gas plug.

2. *Tunnel segments*
   Tunnel rings shall be of precast reinforced concrete of 4.4 m internal diameter, with each ring 600 mm long and consisting of seven ordinary plates, two top plates and one key. Each plate shall be provided with a 50 mm diameter grouting hole and plug.

3. *Bolting cast iron segments*
   The contractor shall provide grummets of approved design under the steel washers to heads and nuts of bolts, to prevent leakage occurring around the bolts. The tightening of bolts shall be carefully performed to ensure that the grummets are forced well home into the bevel of the bolt-holes and that the washers have an even bearing. Where the rings are erected under compressed air, the compressed air equipment shall be retained in working order until the engineer is satisfied that the cast iron lining is watertight.

4. *Caulking longitudinal joints of cast iron segments*
   Caulking of longitudinal joints shall be carried out as soon as practicable after the cast iron lining is erected. Prior to caulking, the recesses shall be thoroughly cleaned by air jets, water jets, scraping or by a combination of these. Metallic lead shall be used for caulking, of the same width as the width of the recess. Where leakage occurs after caulking, the lead shall be removed and renewed. The remaining parts of the caulking recesses shall be tightly filled and pointed with cement mortar (1:3).

5. *Caulking circumferential joints of cast iron segments*
   Preliminary jointing shall be carried out with tarred yarn during erection of the rings. As soon as practicable after erection, the spun yarn packing shall be removed, the joints shall be cleaned as previously described, and a continuous caulking of metallic lead applied behind the bolts, with lead block joints connecting the circumferential and longitudinal joints. The recesses shall be finished off with cement mortar (1:3) after the lead caulking has been checked for watertightness. The contractor's prices for caulking circumferential joints shall include for taking out and replacing bolts, providing, cutting out and removing temporary tarred yarn packings, and cleaning out, caulking and pointing the joints.

6. *Grouting outside cast iron lining*

The space between the shaft lining and the surrounding ground shall be filled completely with cement grout (1:2), mixed with sufficient water to permit it to be forced by compressed air through holes in the castings. The grouting should be carried out as soon as practicable after the lining is erected and the joints caulked, and the holes shall be carefully plugged after grouting.

7. *Jointing precast concrete segments*

The segments of precast concrete rings shall be bolted together with mild steel bolts, domed steel washers and approved grummets. In addition, approved filling compound of 3 mm compressed thickness shall be placed in the circumferential joints, and approved bitumen filling in the longitudinal joints.

The caulking groove between the precast concrete segments shall be thoroughly cleaned, wetted and tightly filled with stiff cement mortar (1:3). The finished work shall be watertight on completion and any unsatisfactory caulking or grummetting shall be cut out and replaced at the contractor's expense.

8. *Grouting outside precast concrete lining*

Cement grout, as previously specified, shall be injected under pressure through the holes in the linings, so as to fill completely all the voids between the tunnel lining and the surrounding ground, and the holes shall be carefully plugged after grouting. Grouting shall commence at the invert and proceed upwards, allowing air and moisture to escape through the upper grout holes.

9. *Concrete lining to shaft and tunnel rings*

A concrete lining of the thickness shown on the drawings shall be applied to the inner face of shaft and tunnel rings, and finished to produce a dense, hard and smooth surface. The concrete shall consist of one part of sulphate-resisting cement to not more than four parts of total dry aggregate by weight, with a maximum aggregate size of 10 mm and a water/cement ratio not exceeding 0.50. The concrete shall have a nominal strength of 35 MN/m^2 and a minimum batch cube strength (6 cubes) in accordance with BS 1881 of 33 MN/m^2 at 28 days. The compacting factor shall not exceed 0.87 and the slump shall not exceed 75 mm.

The concrete shall be adequately mechanically vibrated between steel shuttering of approved design. Fabric reinforcement of type B385 to BS 4483 shall be provided where shown on the drawings.

Source: I.H. Seeley. *Civil Engineering Specification* (Macmillan).

(4) Bill of Quantities

The Civil Engineering Standard Method of Measurement[4] (CESMM) defines a bill of quantities as a list of items giving brief identifying descriptions and estimated quantities of the work comprised in a contract. An explanation of its contents and applications are given in *Civil Engineering Quantities*.[5] The bill of quantities enables all contractors to tender on the basis of the same information.

A bill of quantities contains the following components.

    (a)   list of principal quantities;
    (b)   preamble to the bill of quantities;
    (c)   daywork schedule, if required;
    (d)   work items, usually classified as in the CESMM; and
    (e)   grand summary.

Each of the components is now further considered.

    (a)   The list of principal quantities enables contractors tendering for the work to make a quick assessment of the scope of the contract.
    (b)   The preamble draws the attention of contractors to the method of measurement used in preparing the bill of quantities, together with any amendments which have been made to the method of measurement. Ideally the CESMM[4] should be used without amendments. For highway contracts, the Department of Transport method of measurement for road and bridge works[6] is often used.
    (c)   Where work cannot be adequately defined at the bill preparation stage, the best method is to value it on a daywork basis. One approach is to list craft operatives and labourers and the principal items of materials and plant, with a provisional quantity inserted against each of them. Each item is then priced by the contractor and he will be paid at the rates in the schedule for the actual work done and plant and materials used. Alternatively, and more commonly, the rates contained in the *Schedule of Dayworks carried out incidental to Contract Work*[7] may be used. These rates may be adjusted in accordance with the percentage addition or deduction stated by the contractor in the schedule.
    (d)   It is usual for a bill of quantities to be subdivided into separate parts or bills relating to the various sections of work. For example, a bill of quantities for a roadworks contract might consist of the following bills.

Bill	*Work*
1	General items (as described in class A of the CESMM)
2	Earthworks
3	Roadworks, footpaths and verges
4	Bridges
5	Road markings
6	Road signs
7	Culverts
8	Surface water drainage
9	Street lighting

(e) A Grand Summary at the end of the bill of quantities tabulates the totals of each bill. In addition, provision is normally made for a contingency item to be added to the amounts brought forward from the various bills. Provision is also made for an adjustment item which follows the contingency item, whereby the contractor can adjust his tender figure if he wishes to do so, after full consideration of the work involved, the tendering climate and any other relevant factors. The total of these items constitutes the contractor's tender for the contract.

The valuation of variations and securing agreement to the cost of delays have generally been the main causes of disputes on civil engineering contracts, stemming largely from variations in design and unforeseen physical conditions and artificial obstructions. Furthermore, many of the costs arising from civil engineering operations are not proportional to the quantity of the resulting permanent work. *The Civil Engineering Standard Method of Measurement*,[4] in accordance with which many civil engineering bills of quantities are prepared, endeavours to remove these inconsistencies by the introduction of method-related charges, which are of two basic types: time-related charges and fixed charges. For example, the cost of bringing an item of plant on to the site and its subsequent removal is a fixed-charge and its running cost is a time-related charge.[5]

The contractor can enter and price such costs as he considers he cannot recover through measured rates, such as site accommodation, site services, plant, temporary works, supervision and labour items – all are at the tenderer's discretion. Accepting that expertise in design rests with the engineer, it seems equally evident that expertise in construction methods lies with the contractor. It is logical that the contractor should be able to decide the method of carrying out the works. A blank section in the bill of quantities permits him to list, describe and price these items.

(5) Instructions to Tenderers

Instructions to tenderers aim to assist tenderers in the preparation of their tenders, and to ensure that they are presented in the form required by the employer and the engineer. These instructions will vary from project to project, but some of the important items usually included are as follows:

(1) any documents that are to be submitted with the tenders, such as the programme and mode of procedure, and details of the legal and financial status and technical experience of the tenderer;
(2) the place, date and time for the delivery of tenders;
(3) instructions concerning the visiting of the site by the tenderer;
(4) instructions on whether tenders based on alternative designs will be considered and, if so, the conditions under which they may be submitted;
(5) notes drawing attention to any special conditons of contract, materials and methods of construction, and unusual site conditions; and
(6) instructions relating to the completion of the bill of quantities, submission of tender and supply of performance bond.[1]

Where tenders are invited, often from a selected list as described in chapter 3, particulars of the tenderer's legal and financial status and technical experience are required at an earlier stage.

(6) Form of Tender

The form of tender is the tenderer's written offer to carry out the work in accordance with the other contract documents. It incorporates the total tender sum, the time for completion and other matters pertaining to the offer. Normally the tenderer submits a tender complying fully with the specification, but in some instances he is permitted to offer alternative forms of construction. The employer's written acceptance of the offer is binding, pending the completion of the agreement.

In addition, the appendix to the form of tender has to be completed by the engineer or the contractor. The principal items incorporated in the appendix are now listed and described.

(i)   Amount of bond: this is normally calculated at 10 per cent of the tender total, but no bond may be required where the employer is fully satisfied as to the contractor's financial and legal status.
(ii)   Minimum amount of insurance: this is frequently calculated at 25 per cent of the tender total or £500 000, whichever is the greater.
(iii)   Time for completion: this may be inserted by either the engineer or the contractor; where prescribed by the engineer it simplifies the task of comparing tenders. Provision can also be made for inserting the times for completion of various sections of the works.

(iv) Liquidated damages for delay: this constitutes the employer's genuine estimate of the damages he will sustain if the works are not completed on time. They usually cover interest on capital, cost of prolonged site supervision and loss of revenue from the works.

(v) Period of maintenance: often 12 months.

(vi) Vesting of materials not on site: goods and materials are listed for which payment will be made prior to their delivery to the site.

(vii) Standard Method of Measurement adopted in the preparation of bill of quantities: this is normally the Civil Engineering Standard Method of Measurement.[4] If some other method has been used, details have to be inserted.

(viii) Percentage for adjustment of prime cost sums: the percentage is inserted by the contractor and is then applied to all prime cost sums in the bill of quantities as and when they are adjusted.

(ix) Percentage of the value of goods and materials to be included in interim certificates: often 80 per cent.

(x) Minimum amount of interim certificates: this may be calculated at about one-half of the engineer's estimated average monthly value of the contract. For example, if the engineer's estimate of the contract is £1.5m and the contract period is 12 months, then the average monthly value of work is £125 000 and the minimum amount of interim certificates could thus be £62 500.

(7) Form of Agreement

The form of agreement is also incorporated in the ICE Conditions and constitutes a legal undertaking between the employer and the contractor for the construction of the works, both permanent and temporary, in accordance with the contract documents, and for the maintenance of the permanent works. The employer covenants to pay the contractor at the times and in the manner prescribed by the contract.

(8) Form of Bond

The performance bond is a document whereby a bank, insurance company or other acceptable guarantor undertakes to pay a specified sum if the contractor fails to discharge his obligations satisfactorily. The amount of the bond is usually 10 per cent of the tender sum and the contractor is almost certain to include the cost of providing the bond in his tender. Hence, where the contractor on investigation is found to be in all respects satisfactory, the bond requirement can be omitted and the employer saved the additional costs involved.

THE ICE CONDITIONS OF CONTRACT

Introduction

The first standard set of ICE Conditions was issued in 1945 and the current fifth edition was published in 1973 and revised in January 1979, and reprinted with some minor amendments in January 1986. The form of contract is issued together with the form of tender, agreement and bond, as described earlier in the chapter. It exists in one form only and, unlike the JCT form for building contracts, there are no separate editions for public and private use. Throughout this chapter the use of capital letters in such words as contractor, engineer, employer and works have been omitted to secure uniformity throughout the book, even although this conflicts with the procedure adopted in the Conditions of Contract.

The ICE Conditions create a remeasurement or measure and value type of contractual arrangement. Thus the contractor is paid at the contract rates for the actual quantities of work completed. There is no reference in the Conditions to the term 'contract sum', but instead 'tender total' is used and this is defined as the total of the priced bill of quantities at the date of acceptance of the contractor's tender for the works. The 'contract price' is defined as the sum to be ascertained and paid in accordance with the provisions contained in the Conditions for the construction, completion and maintenance of the works in accordance with the contract.

The ICE Conditions provide a comprehensive contractual code embracing all work and practices normally encountered in civil engineering construction. The engineer has much wider powers and duties than does an architect under the JCT contract. For example, the engineer has power to give instructions as to the method of carrying out the work and he has the duty of adjusting, where necessary, the sums payable to the contractor.

Some argue that the current ICE Conditions are over-elaborate and sometimes vague. However, irrespective of the rights or wrongs of these criticisms, the Conditions do permit a much wider use of the legal concept of reasonableness in the implementation of the Conditions, and the powers and decisions of the engineer. Under the Conditions, the works are to be carried out to the satisfaction of the engineer. Thus he has powers of control and direction, subject to his legal responsibilities as agent of the employer, while, at the same time, being required to act reasonably in accordance with the law.

The Contract

The Conditions of Contract recognise that the form of contract will be accompanied by drawings, specification, bill of quantities, tender, written acceptance of the tender and the contract agreement. These documents are mutually explanatory of one another. If any matter requires clarification, the engineer has the right to make a ruling, although he is required to issue to the contractor such instructions in writing as he considers necessary to enable the contractor to carry out the contract.

The engineer, although possessing wide powers of interpretation, is not permitted to alter the conditions agreed upon at the outset. For example, he has authority to supply the contractor with modifications to drawings and instructions during the progress of the work 'as shall in the engineer's opinion be necessary for the purpose of the proper and adequate construction, completion and maintenance of the works and the contractor shall carry out and be bound by the same' (sub-clause 7.2).

Engineer's Representative

The engineer is required to notify the contractor in writing of the appointment by him or by the employer of an engineer's representative, commonly known as a resident engineer. The functions of the engineer's representative are to watch and supervise the construction, completion and maintenance of the works (sub-clause 2.1).

The generality of this basic function is qualified by the express prohibition on the ordering of any work involving delay or any extra payment by the employer or the making of any variation. The contractor would be wise to refer any questionable matters to the engineer for written confirmation. Acquiescence in acts of the engineer's representative which are *ultra vires* and which result in additional expense and/or delay could be a bar to compensation. In like manner, the engineer's representative should seek the engineer's guidance where there is doubt in such matters as standards of workmanship, materials or testing.[8]

General Obligations

Contractor's General Responsibilities

These are covered in clause 8 whereby the contractor is made responsible for all aspects of the construction, completion and maintenance of the works. This includes the provision of all labour, materials, transport and everything required for the construction, completion and maintenance, so far as the necessity for providing the same is specified in, or is reasonably to be inferred from, the contract. This clause also makes the contractor fully responsible for the adequacy, stability and safety of all site operations and methods of construction, but excluding the design or specification of the permanent works. The safety of the works during construction and, in particular, the temporary works is of paramount importance. It is reasonable to expect the responsibility for this to lie with the contractor, except where any temporary works are designed by the engineer.

Inspection of Site

Under clause 11, the contractor is deemed to have inspected and examined the site and its surroundings, including the ground and sub-soil conditions, before submitting his tender. He is also responsible for ensuring suitable access and means of communication, and in general to have obtained all necessary information as to risks and contingencies. The contractor is entitled to take into account any information with regard to the nature of the ground and sub-soil conditions provided by, or on behalf of, the employer.[9]

Sureties

Where a bond is required from the contractor when submitting his tender for the contract, his sureties, usually amounting to 10 per cent of the tender total, must be banks, insurance companies or other institutions acceptable to, and approved by, the employer.

Adverse Physical Conditions

Clause 12 makes provision for contingencies arising from physical conditions, not attributable to the weather, or artificial obstructions which could not reasonably have been foreseen by an experienced contractor. This latter criterion is very much dependent upon the type of work and what can be deduced from site information and local knowledge. The contractor, if he intends to make a claim for extra payment for additional works or interference with the approved programme, must keep adequate records and, as soon as possible, give notice to the engineer of the problems encountered and, with the notice, or as soon as practicable thereafter, give details of the anticipated effects of the measures he is taking, or proposes to take, and the extent of the anticipated delay or interference with the execution of the works.

The engineer, if he thinks fit, may order a suspension of the works under clause 40 or a variation under clause 51. He may, on the other hand, require the contractor to provide an estimate of cost or give written instructions as to how the physical conditions or artificial obstructions are to be dealt with. In this context an artificial obstruction could be buried pipework or other services encountered during excavation.

Programme of Work

Clause 14 requires the contractor to supply the engineer with a programme of work within 21 days of the acceptance of his tender. The contractor is also required to furnish the engineer with a general description of the arrangements and methods of construction which he proposes to adopt. The engineer, in turn, must provide the contractor with written evidence of his consent to the contractor's proposals or to specify in what respects they fail to meet the contract requirements. There could be a conflict between considerations of margins of safety by the engineer and the profit motive of the contractor.[10]

The engineer should ensure that the contractor is supplied with sufficient information to enable the programme to be prepared. An IQS report[8] suggests that this information should include an indication of the probable date for commencement. This date has to be notified by the engineer in writing within a reasonable time after the acceptance of the tender (clause 41).

In the absence of any provisions to the contrary elsewhere in the contract, the contractor is not required to produce a costly critical path network.[8] Responsibility for the methods of construction adopted is recognised to be a matter for the contractor, and the engineer's primary responsibility is to satisfy himself that they are such as to enable the work to be executed without detriment to the permanent works.[10]

Contractor's Superintendence

The contractor shall provide all necessary superintendence during the execution of the works by sufficient persons having adequate knowledge of the operations to be carried out (clause 15). Written approval by the engineer to the contractor's authorised agent or representative is required, and this approval may be withdrawn at any time. The authorised agent or representative of the contractor is to be constantly on the works and is required to give his whole time thereto. It is important to note that the contractor or his agent or representative is responsible for the safety of all operations.

Removal of Contractor's Employees

Clause 16 gives the engineer wide powers to object to the employment on the works of any of the contractor's employees, and to require the removal of any person who misconducts himself, is incompetent, is negligent in the performance of his duties, fails to conform to any particular provisions with regard to safety, or who persists in any conduct which is prejudicial to safety or health.

Setting Out

Clause 17 makes the contractor responsible for the true and proper setting out of the works and for the correctness of the position, levels, dimensions and alignment of all parts of the works. The checking of any setting out by the engineer, or the engineer's representative, shall not relieve the contractor of his responsibilities.

Only if incorrect data is given by the engineer or the engineer's representative, can the contractor claim for reimbursement of the cost of rectifying errors.[9] Having regard to the heavy responsibility of the contractor under this clause, the engineer shall ensure that the contractor is in possession of full and accurate information for the work concerned and for any instructed variations.[8]

Insurance

This complex area is covered in clauses 20 to 25, and readers wishing to examine these provisions in detail are referred to *Insurance under the ICE Contract*.[11] Clause 21 imposes responsibility on the contractor for loss or damage to the works except in regard to the excepted risks, when the cost of making good is borne by the employer. The 'excepted risks' are riot, war and similar events, and any cause due solely to the engineer's design. The contractor is required to insure jointly in the names of the employer and the contractor against any loss or damage for which he is responsible, and to submit the name of the insurer and the terms of the insurance to the employer for his approval.

Clause 24 provides for insurance against accidents to workers, and specifically requires this insurance as a protection for the employer, since injured workmen have, on occasions, sued the employer rather than the contractor. Hence the employer is indemnified by this clause.

Under clause 25, if the contractor fails to produce to the employer, upon request, satisfactory evidence to show that any insurances required to be taken out by the contractor are in force, then the employer may himself effect any such insurances and deduct the premium paid by him from payments otherwise due, or which may become due to the contractor, or recover the amount concerned as a debt due from the contractor. The contractor should ensure that he is in a position to supply, whenever required by the employer, documentary evidence in the form of policies and receipts for current premiums to show that he is fulfilling his contractual obligations.

Giving of Notices and Payment of Fees

Sub-clause 26(1) requires the contractor to give all notices and pay all fees required to be given and paid by any Act of Parliament, regulation or bye-law of any local or statutory authority, in relation to the execution of the works or temporary work, and by the rules and regulations of all public bodies and companies whose property or rights may be affected by the works or temporary works.

The engineer shall certify the sums which have been properly paid by the contractor in compliance with this clause together with all rates and taxes paid by the contractor in respect of the site or any part thereof, or anything constructed or erected on it, or any temporary structures situated elsewhere but used exclusively for the purposes of the works. Such sums are to be paid or allowed to the contractor, normally through interim certificates.

Contractor to Conform with Statutes (sub-clause 26(2))

The contractor must ascertain and comply in all respects with the provisions rules and regulations described in sub-clause 26(1), and is to keep the employer indemnified against all penalties and liabilities of every kind for breach of any such act, regulation or bye-law. The contractor should bring to the attention of the engineer, any failure of the drawings, specification or instructions to conform with any act, regulation or bye-law, and the engineer should immediately investigate the matter and supply the contractor with modified drawings, specification or instructions. If this results in the contractor incurring delay or cost, then such delay should be taken into account by the engineer when determining the amount of any extension of time and the contractor should be paid such costs as may be fairly and reasonably attributable thereto.

The employer's warranty relating to planning permission covers the permanent works and any temporary works specified or designed by the engineer. The contractor should satisfy himself with regard to any further planning permission which may be required in connection with separate proposals by the contractor as, for example, off-site offices and storage areas required for his own purposes.[8]

Avoidance of Damage to Highways

Under clause 30 the contractor should, insofar as is practicable, ascertain, prior to submitting his tender, whether any of the routes selected are likely to be subject to restrictions by interested authorities, and he and his sub-contractors should, as far as practicable, select such routes and use vehicles and distribute loads and take all other reasonable means to reduce the likelihood of roads being subjected to extraordinary traffic.

Unless the contract provides otherwise, the contractor is responsible for and is to pay the costs of strengthening any bridge or altering or improving any highway, communicating with the site, for the purpose of transporting constructional plant, equipment or temporary works, and is to indemnify the employer against such claims. Where this work results from a variation the contractor should be repaid the costs incurred by him under this clause.

The employer is to indemnify the contractor against claims for damage caused to bridges or highways due to the transporting of materials or manufactured or prefabricated articles required in the execution of the works, unless it arises from a failure by the contractor to fulfil his obligations as previously described under this clause.

Facilities for Other Contractors

The contractor is to afford all reasonable facilities to other relevant persons or bodies as required by the engineer. It is advisable for the engineer to include in the tender documents full details of the works to be executed by other contractors, such as the extent, timing and duration of these works, and the amount of notice required where they are to be co-ordinated with the main contractor's programme.[9] The contractor will appreciate that he is not entitled to exclusive possession of the site and should endeavour to ensure that he does not unnecessarily obstruct or interfere with the operations of others, and the employer owes a similar obligation to the main contractor in respect of the works of other contractors.

If the contractor is delayed or incurs additional costs in complying with clause 31 beyond what reasonably could be foreseen by an experienced contractor, the engineer must take account of such delay when determining the contractor's entitlement to extension of time (clause 44) and any additional payment (clause 60).[8]

Clearance of Site on Completion

The contractor is required to clear all plant, surplus material, rubbish and temporary works from the site on completion of the permanent works under clause 33. However, the contractor may wish to retain certain items of plant on the site for remedying defective work, subject to the overriding obligation that it must not be detrimental to the beneficial use of the works by the employer.

Workmanship and Materials

Cost of Samples and Tests

Where tests are required by the engineer under clause 36 these should be notified to the contractor in writing in sufficient time to permit the contractor to arrange for a representative to be present. The engineer should confine the purpose of his selecting and testing of samples to determining their compliance with the contract.

Unless the supply of materials or carrying out of tests is clearly intended by, or provided for, in the contract, these are to be at the cost of the employer. However, the contractor will be responsible for the cost of making a test if it shows the workmanship or materials not to be in accordance with the contract or the engineer's instructions.

Examination of Work before Covering Up

Under clause 38 the contractor must notify the engineer, preferably in writing, whenever any work is ready for examination by the engineer before being covered up. No such work is to be covered up without the approval of the engineer, except where the engineer advises the contractor that he considers an inspection is unnecessary. The engineer is required to carry out his inspection and measurement without unreasonable delay.

The engineer is empowered to order the contractor to uncover or make openings in or through any part of the works and to reinstate them to the satisfaction of the engineer. The expenses incurred are to be borne by the contractor unless he has complied with the conditions described in the previous paragraph and the work is found to have been executed in accordance with the contract, in which case they are to be borne by the employer.

Removal of Improper Work and Materials

The engineer is empowered under clause 39 to order the contractor to remove materials which are not in accordance with the contract and to substitute proper and suitable materials, and to similarly remove and replace any unsatisfactory work.

Suspension of Work

Clause 40 gives the engineer power to suspend the progress of the whole or part of the works. If the suspension lasts for more than three months, the contractor may serve notice on the engineer requiring further permission to proceed with the works within 28 days. In the extreme, if permission is not granted within a further 28 days, the contractor may treat this as an abandonment of the contract by the employer.

Commencement Time and Delays

Commencement of Works

Under clause 41 the date for commencement must be within a reasonable time after acceptence of the tender, and the time for completion is calculated from the date for commencement. If the contractor delays commencement or does not proceed with due diligence, and the engineer is of the opinion that such delay or lack of progress is unreasonable and unwarranted, then he should give consideration to the provisions of clause 46 or, more seriously, those of clause 63 (forfeiture).

Possession of Site

Under clause 42 the employer is to make available on the date for commencement so much of the site as will enable the contractor to commence and proceed in accordance with the programme submitted under clause 14, and thereafter make available such further portions as will enable him to adhere thereto, otherwise the contractor will be entitled to claim reimbursement of the additional costs incurred.

Extension of Time for Completion

Clause 44 stipulates the various causes which give rise to an extension of time, and requires the contractor to give 28 days notice of these circumstances.

Rate of Progress

The engineer is required under clause 46 to notify the contractor in writing if he considers the rate of progress too slow to ensure completion on time. The contractor must then take steps to expedite the work. If night or Sunday working is proposed, consent should not be unreasonably withheld by the engineer. Night work must be carried out without causing unreasonable noise and the contractor must indemnify the employer against damages resulting from noise and disturbance. Liquidated damages for delayed completion are dealt with in chapter 8.

Completion and Maintenance

Completion Certificate

Under clause 48 the contractor is liable up to the issue of the completion certificate for the work which he has contracted to carry out. Thereafter, the contractor becomes subject to the maintenance provisions of the contract, and must repair any defects and maintain the work for the agreed period of maintenance.

The contractor may request the engineer to test, approve and certify the works, and, at the same time gives an undertaking to finish any outstanding work during the maintenance period. The engineer has 21 days from receiving notice of the request in writing from the contractor seeking a certificate of completion, to issue it to the contractor.

Maintenance and Defects

Clause 49 deals more specifically with the maintenance details and with action to be taken on defects. The contractor is obliged under this clause to rectify any defects whether or not they are due to his failure to comply with the contract, excluding fair wear and tear. If it is determined that the defects are not due to failure by the contractor then he is entitled to claim payment. In default, the engineer may arrange for the work to be carried out by his own workmen or by other contractors, and to recover the cost from the contractor or to deduct it from any monies due to him.

Alterations, Additions and Omissions

Variations

Under clause 51 the engineer's powers include varying the works and also order-ing changes in the specified sequence. The engineer binds the employer by any variation he orders, but the engineer himself is not bound to vary the work on the request of the contractor. It is significant to note that such variations do not vary the contract itself by the work undertaken under the contract, or reasonably consequential upon it.

The contractor should not make any variation unless he has received an order in writing or has received a verbal order subsequently confirmed by the engineer or contractor (the latter not having been contradicted in writing by the engineer). Should the engineer fail to confirm his oral instruction to vary the works, or contradicts a written confirmation already issued to the contractor, then a dispute exists which is referable to arbitration clause 66. These clauses encompass a wide variety of circumstances which result in variations to the works, ranging from adverse physical conditions to delays occasioned by the finding of fossils on the site. The engineer is not, however, required to issue an order in writing for fluctuations in quantities from those contained in the bill of quantities.

Valuation of Variations

The engineer is required to consult with the contractor prior to determining the value of variations. The value is to be ascertained in accordance with the follow-ing principles.

(1) Where work is of a similar character and executed under similar conditions to work priced in the bill of quantities, then it shall be priced at such rates and prices contained in the bill of quantities as may be applicable.
(2) Where the work is not of a similar character, or is not executed under similar conditions, then the rates and prices in the bill of quantities shall be used as the basis for valuation so far as may be reasonable. If it would not be reasonable, then a fair valuation is to be made.

It is contemplated that evaluation should be arrived at by agreement between the contractor and the engineer within the framework of these principles. In the event of agreement not being reached, the engineer is required to determine the rate or price, in accordance with the established principles, and to notify the contractor accordingly.

Either the engineer or the contractor may give notice to the other that the contract rates or prices should be varied. Such notice must be given either before the commencement of the varied work or as soon as is practicable thereafter. Where such a notice has been given, the engineer is required to fix such rate or price as in the circumstances he shall consider reasonable and proper. In forming his opinion as to whether or not any rates or prices have been rendered unreasonable, and in fixing such other rates or prices as in the circumstances the engineer considers reasonable or proper, it has been suggested that the engineer should have due regard to the component elements of the original rates or prices.[8] The practical application of the valuation of variations will be examined in chapter 7.

Daywork

The engineer may, if he considers it necessary or desirable, order in writing that any additional or substituted work shall be executed on a daywork basis. The following procedure is then followed.

(1) The contractor is required to submit to the engineer for his approval quotations for materials before ordering them.
(2) The contractor is also required to submit each day unpriced particulars in duplicate of the labour, materials and plant involved.
(3) When the particulars are agreed, the engineer's representative should sign the lists and return one copy to the contractor.
(4) If any particulars are found to be incorrect, the engineer's representative should notify the contractor of the reason therefor in order that any adjustments may be agreed.
(5) At the end of each month the contractor is required to deliver to the engineer's representative a priced statement of the labour, materials and plant used. The priced statement is normally prepared in a summarised form based on the detailed lists and statements previously submitted.
(6) The contractor will be paid for this work under the conditions and at the rates and prices contained in the daywork schedule included in the bill of quantities. In the absence of a daywork schedule, the contractor will be paid in accordance with the *Schedule of Dayworks carried out incidental to Contract Work,*[7] current at the date the work is carried out.

Notice of Claims

Clause 52 deals with the notification of, and the procedures to be adopted with regard to claims for additional payment, to which the contractor considers himself entitled under the contract, and the operational aspects are considered in detail in chapter 8.

If the engineer notifies the contractor of a change in the quantity of any item, the contractor must, if he intends to claim a higher rate or price than that notified, give notice within 28 days in writing to the engineer of his intention.

If the contractor intends to claim any additional payment in any other circumstances, then:

(1) The contractor must give notice in writing to the engineer as soon as reasonably possible after the happening of the events giving rise to the claim.
(2) As soon as the events occur, the contractor shall keep such contemporary records as may be reasonably necessary to support any such claim. The contractor would be well advised to obtain the engineer's agreement to the records wherever possible.

As soon as is reasonable, the contractor is required to send to the engineer:

(1) A first interim account giving full and detailed particulars of the amount claimed at that date, and of the grounds on which the claim is based. This should be, at the very latest, with the interim monthly statements, updated in subsequent months.
(2) When required by the engineer, to send further updated accounts and details. The timing should be such as to enable the engineer to give full consideration to any such claims, and also to allow interim payments to be certified either in whole or in part.[9]

Property in Materials and Plant

Plant

Clause 53 provides that plant and materials brought on to the site and owned by the contractor are to be legally vested in the employer. Certain types of plant under hire-purchase agreements, where removal from the site would endanger the safety of the structure or cause a serious disturbance of the works, must be contracted for so as to safeguard the position of the employer, should the contractor fail, or be in dispute with the owning company. Thus, for example, the employer can be included in hire-purchase contracts made by the contractor, so as to allow him to take over overdue hire-purchase payments on plant, to prevent its repossession by the plant hirer. Any request by the employer to take over the hire of plant must be made, in writing, within seven days of the date of forfeiture. The engineer is entitled, on request, to be notified in writing of the name and address of the owner of any specific item of plant.

Action by Contractor

The contractor must obtain the engineer's written consent before removing constructional plant, goods and materials from the site. However, the engineer must not unreasonably withhold his consent to the removal from the site of plant, goods and materials not required immediately for the purpose of completing the works. Until the occurrence of any event entitling the employer to exclude him from the site, the contractor is entitled to the exclusive use, for the purpose of completing the works, of all the items of plant, goods and materials brought on to the site by the contractor, which are deemed to have become the property of the employer.

The property in plant, goods and materials removed from the site by the contractor, with written consent, is immediately revested in the contractor. Upon completion of the works, the property in plant, goods and materials remaining on the site is revested, subject to the provisions of clause 63, in the contractor.

After completion of the works, the contractor must remove all plant, goods and materials as required, within such reasonable time as may be allowed by the engineer. In the event of the contractor failing to remove any such items, the employer is entitled to sell any that are owned by the contractor and to return to the owner, at the contractor's expense, any items remaining on the site which are not the property of the contractor.

The employer is entitled to deduct from the proceeds of any such sale any costs he may have incurred in connection with the sale and/or the return of equipment. The contractor is entitled to the balance of the proceeds, but in the event of the sale proceeds failing to cover the expenses incurred by the employer, the contractor shall be liable to the employer for any balance outstanding. It is advisable that the engineer affords the contractor adequate time to arrange for the removal of plant, taking into account all material circumstances including maintenance obligations, and that he obtains the contractor's proposals before deciding the length of a reasonable period.[8]

Vesting in Employer

Upon the engineer's approval in writing of goods and materials not on the site, they will vest in and become the absolute property of the employer. Notwithstanding such vesting in the employer, the contractor will be responsible for any loss or damage, cost of storage, handling and transporting of such goods and materials. In addition, the contractor is required to effect any additional insurance necessary to cover the risk of such loss or damage arising from any cause.

Measurement

Quantities

Clause 55 provides that the quantities set out in the bill of quantities are the estimated quantities of the work, but they are not to be taken as the actual and correct quantities of the work to be executed by the contractor in fulfilment of his obligations under the contract, indicating that total remeasurement of the works in inevitable. Furthermore, any error in description in, or omission from, the bill of quantities shall be corrected and the value of the work actually executed is to be ascertained in accordance with clause 52, as described previously.

Measurement and Valuation

Clause 56 contains the following important provisions:

(1) Except as otherwise stated, it is the duty of the engineer to ascertain and determine by measurement the value in accordance with the contract, of work done thereunder.
(2) The engineer is empowered, after consultation with the contractor, to vary rates or prices where they have been rendered unreasonable or inapplicable as a result of fluctuation in quantities. Should the contractor disagree with the engineer's proposals, then the contractor is required to invoke the provisions of sub-clause 52(4)(a).
(3) Where the engineer requires any part of the works to be measured, he must give reasonable notice of his requirements in writing to the contractor, who is required either to attend, or to send a qualified agent to assist in making the measurement.
(4) If the contractor fails to attend or to send a representative, then the measurement made by the engineer or his representative shall be accepted as the correct measurement.
(5) It is essential that any delegation of the engineer's duties under this clause and any limitations placed thereon shall be fully detailed and notified in writing to the contractor.

Method of Measurement

Clause 57 states that except where expressly indicated, bills of quantities shall be deemed to have been prepared in accordance with the procedure set out in the *Civil Engineering Standard Method of Measurement*.[4]

Nominated Sub-contracts

General Arrangements

Under clause 4, the contractor must obtain the engineer's consent to sub-contract any part of the works to someone of his own choosing. The contractor who is given permission remains fully liable for his own choice of sub-contractor. A standard form of sub-contract for use in conjunction with the ICE form of contract is available.

Clauses 58, 59A and 59B stipulate the complex rules and provisions governing the rights of all parties, where the contract provides for specific tasks to be carried out by nominated sub-contractors. Such work is designated as either a provisional sum or a prime cost sum. If the item is prime cost, then the engineer has power to order the contractor to employ a nominated sub-contractor to execute work or to supply goods, materials or services. In the case of provisional sums (sums included in the contract and so designated for the execution of work or the supply of goods, materials or services or for contingencies), such sums may be used in whole or in part, or not at all, at the discretion of the engineer.

The engineer can pay direct to any nominated sub-contractor, if the contractor has unreasonably withheld what is due. The engineer can also hold retention money on sums due to nominated sub-contractors.

Nominated Sub-contractors – Objection to Nomination

Clause 59A allows for objections by contractors to nominees of the employer, and also requires nominated sub-contractors to indemnify the contractor from and against any negligence by the sub-contractor or his workers against any misuse of plant. Despite this, having accepted the nomination of a sub-contractor, the contractor becomes legally responsible for the nominated sub-contractor (for the work executed or goods, materials or services supplied) as if he had himself executed the work or supplied the goods.

Forfeiture of Sub-contract

Clause 59B lays down the procedure to be followed in the event of the contractor being of the opinion that he is entitled to terminate the nominated sub-contract. The contractor is required to notify the engineer in writing at once, if an event arises which the contractor considers entitles him to exercise his right under the forfeiture clause, and he also requires the employer's consent. The contractor is required to take all necessary steps and proceedings to enforce the provisions of the sub-contract and to recover the employer's loss from the sub-contractor.

Certificates and Payment

Monthly Statements

Under clause 60, the contractor is required to submit to the engineer a monthly statement showing the following:

(1) estimated value of the permanent works executed, valued at contract rates;
(2) list of unfixed goods and materials and their value;
(3) list of goods and materials and their value, which are vested in the employer but not yet delivered;
(4) estimated amounts to which the contractor considers himself entitled under the contract, such as temporary works or constructional plant for which separate amounts are included in the bill of quantities;

unless the contractor considers that the total estimated value would fall below the sum inserted in the Appendix to the Form of Tender as being the minimum amount of interim certificates. Amounts payable in respect of nominated sub-contractors are to be listed separately.

Retention

The engineer is required to certify, and the employer is required to pay within 28 days of the delivery of the monthly statement to the engineer or his representative. The payments are subject to retention at 5 per cent of the amount due, excluding materials on site or vested in the employer, subject to the following limits:

(1) 5 per cent for tender totals less than £50 000 with a limit of £1500.
(2) 3 per cent for tender totals in excess of £50 000.

Half of the whole of the retention, less any sums already paid, shall be released within 14 days of the issue of the certificate of completion. The balance of all retention monies, less the amount representing the cost of any works remaining to be executed is released within 14 days of the expiration of the period of maintenance. Failure by the engineer to certify, or the employer to make payment, renders the employer liable to pay interest to the contractor on any overdue payment, calculated at the minimum lending bank rate (base rate) plus ¾ per cent. The GC/Works/1[12] contract for government projects allows for an interim advance midway between monthly valuations when the contract sum exceeds £100 000.

Final Account

The contractor is required to submit a statement of final account, together with all supporting documentation not later than 3 months after the date of the maintenance certificate. The engineer shall issue a final certificate showing the amount finally due to the contractor within 3 months of receipt from the contractor of the final account and of all reasonable information required for its verification. The balance of any monies due from the employer to the contractor or from the contractor to the employer shall be paid within 28 days of the date of the final certificate.

Remedies and Powers

Urgent Repairs

Clause 62 gives the employer the right to carry out any remedial or repair work which in the opinion of the engineer is urgently necessary and which work, or repair, the contractor is either unable or unwilling to do at once. If the liability for this work under the contract is at the contractor's expense, then the contractor shall pay the employer, on demand, all costs and charges properly incurred, or alternatively the employer may deduct such monies from any payments due to the contractor.

Forfeiture

Clause 63 deals with the forfeiture of the contract and sets out the various reasons where forfeiture of the contract by the actions, or lack of actions, of the contractor becomes applicable. In view of the far reaching implications of this clause, which stem essentially from breaches of the contract by the contractor, the engineer should only exercise his power to certify under this clause in circumstances where, after the most careful consideration of all relevant matters, he is of the firm opinion that irrefutable evidence exists of any matters specified in (a) to (e) of sub-clause 63(1). The employer is strongly advised to carefully consider the basis of the engineer's certificate and, in all probability, to take legal advice before giving written notice to the contractor.[8]

Frustration

Clause 64 deals with the frustration of the contract, and both the employer and contractor are strongly advised to obtain legal advice on the question of whether or not the contract has been frustrated in the legal sense.[8]

War Clause

Clause 65 deals with the procedures to be adopted in the event of the outbreak of war, and these are generally self-explanatory.

Settlement of Disputes

Arbitration

The engineer is in a powerful position in that his decision on any disputed matter is binding on the contractor, who must proceed with the works with due diligence and give effect to all decisions of the engineer, unless and until they have been revised by an arbitrator. If the contractor is dissatisfied with the engineer's decision he may, within three months, require that the matter be referred to arbitration.

It is necessary that a dispute or difference should exist before an engineer's decision is sought or given. For example, the initial rejection of a claim or contention does not constitute an engineer's decision. It is desirable that disputes or differences referred to the engineer for decision under clause 66, which deals with the settlement of disputes between the employer and the contractor by recourse to arbitration, are clearly stated to have been submitted for that purpose, as opposed to being an attempt to negotiate a settlement before resorting to the provisions of this clause. In like manner, an engineer's decision should be worded with this intention in mind.

The engineer should state his decision in writing and give notice to both the employer and the contractor within three calendar months of the request for the decision having been made. Should he fail to give a decision within that time then either the employer or the contractor may within a further three calendar months require that the matter be referred to an arbitrator. If, on the other hand, the engineer gives his decision within three calendar months of being requested to do so, then if either the employer or the contractor is dissatisfied with the decision, either party may at any time within three calendar months of being notified of the decision require that the matter be referred to an arbitrator.[8]

In certain circumstances the Courts may extend the time within which a request for reference to arbitration must be made. Such circumstances would have to be very exceptional and it is therefore important that the time limits prescribed in this clause should be observed.

Neither party should make application to the President of the Institution of Civil Engineers for an arbitrator to be nominated unless the parties fail to appoint an arbitrator within one calendar month of either party serving on the other party a written notice to concur in the appointment of an arbitrator. Any such reference to arbitration may be conducted in accordance with the Institution of Civil Engineer's Arbitration Procedure (1973) or any amendment or modification thereof in force at the time of appointment of the arbitrator. The award of an arbitrator is final and binding upon the parties, although an appeal may be made on a point of law.

Summing up, clause 66 provides for settlement of disputes by a three phase procedure in cases where the contractor feels aggrieved. He states his claim firstly to the engineer. Should the engineer reject it then he may be requested to reconsider the matter. The engineer's second decision, if rejected by the contractor, is then subject to arbitration. The ICE arbitration procedures are applied leading up to formal arbitration, to ensure that the engineer and the contractor deal with their dispute in a legally correct manner.

Tax Matters

Tax Fluctuations

Clause 69 sets out the procedure for the recovery of tax fluctuations by the contractor from the employer and vice versa. In all cases the level of taxation included in the rates and prices in the bill of quantities, are those ruling at the date for return of tenders.

The contractor may incorporate similar tax fluctuation provisions in any subcontract placed by him. The contractor is required to give notice of all tax fluctuations, keep contemporary records necessary to determine the amounts of all adjustments, permit the engineer to inspect such records and to submit details of adjustments with the monthly interim accounts.[9]

Value Added Tax

If any uncertainty arises with regard to the charging of VAT, both the employer and the contractor should consult the Customs and Excise Authorities.

Under clause 70, tenders are deemed not to include VAT and any VAT chargeable to the employer is to be paid by him at the same time as the payment for goods and services. If the contractor agrees with the tax payment made, he must issue the employer with an authenticated receipt, but if he disagrees then he must notify the employer in writing, stating the grounds for this disagreement. Disputes about VAT are settled through the Commissioners.[9]

Metrication

Clause 71 deals with the problems associated with changes to metrication in the following manner:

(1) If the contractor, having used his best endeavours, cannot obtain the materials described in the contract, or ordered by the engineer, in the metric or imperial measure required, but can obtain materials approximating to those required and thus avoid delay or additional expense, he is required to notify the engineer in writing.

(2) The engineer shall, if necessary, give effect to any design change which may be required, to avoid delay in the performance of the contractor's other obligations under the contract.

(3) If the contractor defaults under this clause, the cost of delay or expense incurred shall be borne by the contractor.

(4) If the contractor cannot obtain the necessary materials without delay, the engineer must recognise this delay under clause 13.

(5) If alternative materials can be supplied, then the engineer shall issue an order under clause 51 to either supply the alternative materials or make some other variation to avoid the need to supply such materials to the dimensions specified. The contractor may then be entitled to additional payment under clauses 13 and 52.

(6) This clause also applies to nominated sub-contractors.[9]

Contract Price Fluctuations

The contract price fluctuations clause is optional and is normally included in contracts of over two years duration. It consists of a cost indices system of variation of price which superseded the former laborious procedure of calculating the price fluctuations from wage sheets and invoices.

The amount to be added to, or deducted from, the contract price is the nett amount of the increase or decrease in cost to the contractor in carrying out the works. The index figures are compiled by the Department of the Environment[13] and comprise:

(1) the index of the cost of labour in civil engineering construction;
(2) the index of the cost of providing and maintaining constructional plant and equipment; and
(3) the indices of constructional materials prices in respect of aggregates, bricks and clay products generally, cements, cast iron products, coated roadstone for road pavements and bituminous products generally, fuel for plant, timber, reinforcing steel and other metal sections, and fabricated structural steel.

The base index figure is the appropriate final index figure applicable to the date 42 days prior to the date for the return of tenders, while the current index figure applies to the last day of the period to which the certificate relates.

The fluctuations apply to the effective value, as and when included in the monthly statements by the contractor and certified by the engineer, and are subject to retention in accordance with sub-clause 60(4). Materials on site are not included in the effective value. Dayworks or nominated sub-contractors' work are only excluded from the effective value if based on actual cost or current prices.

The increase or decrease in the amounts otherwise payable under clause 60 is calculated by multiplying the effective value by a price fluctuation factor, which is the nett sum of the products obtained by multiplying each of the proportions inserted by the contractor against labour, plant and materials in sub-clauses 4(a), (b) and (c), by a fraction, the numerator of which is the relevant current index figure minus the relevant base index figure, and the denominator of which is the relevant base index figure. Provisional index figures used in the adjustment of interim certificates shall be subsequently recalculated on the basis of the corresponding final index figures.

The simplified formula results in a degree of approximation due to:

(1) the extent to which the proportional factors inserted by the contractor vary from the operative figures on the site;
(2) the extent to which the factors and weightings forming the basis for each index vary from those appertaining to the particular contract; and
(3) the pattern of interim payments which may vary significantly from the the pattern of costs incurred.[9]

OVERSEAS FORMS OF CONTRACT

The Conditions of Contract (International) for Works of Civil Engineering Construction,[14] known as the FIDIC contract are recommended for use for works of civil engineering construction which are the subject of international tender. It has evolved from the ICE Conditions of Contract and is used in many countries throughout the world. There is provision for the insertion of the ruling language in which the contract is to be construed and interpreted.

Some of the ICE clauses are amended or extended to suit local conditions and there is provision for the payment of bonus for early completion. The final section of the contract covers dredging and reclamation works. Readers requiring further information on the operation of the FIDIC Conditions of Contract are referred to The FIDIC Conditions.[15]

Another form of contract available for overseas use is the Overseas (Civil) Conditions of Contract[16] which is closely modelled on the 4th edition of the ICE Conditions, suitably amended to meet local conditions, statutes and regulations.

REFERENCES

1. Institution of Civil Engineers. Civil Engineering Procedure. Third edition (1979)
2. Institution of Civil Engineers, Association of Consulting Engineers and Federation of Civil Engineering Contractors. Conditions of Contract and Forms of Tender, Agreement and Bond for use in connection with Works of Civil Engineering Construction. Fifth edition (June 1973, revised January 1979)
3. I.H. Seeley. Civil Engineering Specification. Macmillan (1976)
4. Institution of Civil Engineers and Federation of Civil Engineering Contractors. Civil Engineering Standard Method of Measurement (1985)
5. I.H. Seeley. Civil Engineering Quantities. Macmillan (1977)
6. Department of Transport. Method of Measurement for Road and Bridge Works. HMSO (1977)
7. Federation of Civil Engineering Contractors. Schedule of Dayworks carried out incidental to Contract Work (1975)
8. Institute of Quantity Surveyors, Civil Engineering Committee. Report on Procedure in connection with the ICE Conditions of Contract (1973)
9. Association of Surveyors in Civil Engineering. The Fifth Edition Explained: Notes for guidance on the ICE Conditions of Contract − Fifth Edition, revised 1979 (1979)
10. Royal Institution of Chartered Surveyors, Quantity Surveyors Civil Engineering Working Party. The ICE Conditions of Contract (1973)

11. F.N. Eaglestone and C. Smyth. *Insurance under the ICE Contract.* Godwin/ Longman (1985)
12. *General Conditions of Government Contracts for Building and Civil Engineering Works GC/Wks/1.* HMSO (November 1973)
13. Department of the Environment. *Monthly Bulletin of Construction Indices (Civil Engineering Works).* HMSO
14. Fedération Internationale des Ingénieurs Conseils and the Fedération Internationale Européenne de la Construction. *The Conditions of Contract (International) for Works of Civil Engineering Construction* (1977)
15. J.G. Sawyer and C.A. Gillott. *The FIDIC Conditions.* Telford (1985)
16. Association of Consulting Engineers and Export Group for the Construction Industries. *The Overseas (Civil) Conditions of Contract* (1956)

3 Estimating and Tendering

This chapter is concerned with tendering procedures and contractor selection, the methods of preparing estimates by contractors and the compiling of tenders, the appraisal of tenders by the engineer, the methods used for accepting tenders and notifying the results, and finally the signing of the contract.

BROAD AIMS

The principal aims of all members of the construction team should be to complete the project to the required quality and form, within the planned time and at the optimum cost, thus providing value for money. To achieve these objectives a CIOB working party[1] advocated the following general guidelines.

(1) Selection of a management system for both design and construction which is best suited to the employer's and the project's needs.
(2) Provision of a clear brief setting out the employer's requirements.
(3) Deciding the stage to which design shall have advanced before work starts on site. This involves consideration of the time/cost factors in the design and construction process, and an understanding of the relationship of the return obtained either as income, amenity or other benefit from the investment.
(4) Having taken into account (1) and (3), to decide the appropriate basis on which a contractor shall be selected.
(5) Appoint a suitable contractor and use him effectively.
(6) Ensure that the commercial risks to the employer and the contractor are identified, fairly apportioned and allowed for when selecting the contractor, and in the tendering arrangements and contract documents.

TENDERING PROCEDURES

Open Tendering

This is a method occasionally used by public authorities to obtain tenders by advertising in the press and/or technical journals, and any contractor who responds to the advertisement is supplied with the tender documents. A deposit is usually requested which is returned on receipt of a *bona fide* tender. The employer does not bind himself to accept the lowest or any tender. Ideally, tenderers should be informed, at the time of issuing the tender documents, of the number of contractors to whom issue has been made.

This procedure offers the principal advantage of full competition from all interested contractors. It can, however, result in an excessively long list of tenderers with much abortive tendering and waste of resources. It has been estimated that tender costs for civil engineering projects average ½ per cent of company turnover. It could also result in tenders being awarded to contractors who are not adequately equipped financially or technically to undertake the work. It is a very inefficient process and is now little used.

The Wood Report[2] advised against its use and stated 'We cannot, however, endorse the use of open competition. It has little to offer over some form of selection prior to invitation to tender, and we cannot entertain any justification for its continued use in the face of repeated condemnation in past reports, and the poor performance on such contracts in our statistical survey'.

Selective Tendering

Selective tendering, based on approved lists or pre-qualification, is strongly recommended in an ICE document[3] as being the best procedure, affording maximum efficiency and economic advantage. It offers both a popular and a relatively straightforward procedure, ensuring the receipt of meaningful tenders with the least delay.

In this method tenders are invited from selected contractors chosen by the engineer and the employer. Lists of contractors suitable for specific categories and values of civil engineering projects are compiled by both large promoting organisations and consulting civil engineers. Only contracting organisations of repute with adequate financial and technical resources will be listed, and the lists are normally subject to periodic review.

A NJCC Code[4] recommends that a short list of suitable tenderers should be drawn up either from the employer's approved list of contractors or from an *ad hoc* list of contractors of established skill, integrity, responsibility and proven competence for work of the character and size contemplated. Although this code is aimed at building contracts its provisions are equally applicable to civil engineering work.

Another ICE publication[5] describes how the practice of inviting tenders from a list of selected contractors has the principal advantage of eliminating the worst features of open tendering. It does however make it more difficult for reputable contractors to secure the opportunity of tendering in a new field. The compilation and periodic review of lists of contractors to be invited for specific types and sizes of projects requires a detailed knowledge of their respective achievements, capabilities and reputations.

Two-stage Tendering

Two-stage tendering is of value where early contractor selection is required but it is not feasible to negotiate with a single contractor without any form of competition. The first stage involves the competitive selection of the contractor, while the second stage embraces the determination of the contract price based on pricing data obtained from the first stage. Sufficient information will be supplied to tendering contractors to enable them to establish the basis on which the final price will be determined. Thus the offers of selected contractors may be considered in the light of such factors as management and plant capacity, labour rates and overheads, and they may even be asked to price a nominal bill. In some instances the first stage is preceded by a preliminary stage in which the selected contractors are interviewed to determine the extent of their resources and the contribution they can make. It can also be a subjective assessment leading to the first stage tender.

In the second stage after the design has been prepared, with the contractor participating in the design team, the contract price is determined. At this stage it is very important that the determination of the actual price is closely related to the quantity and specification of the work to be done, and that the cost control mechanism for use during construction is established.

The main advantages to be gained from two-stage tendering are:

(1) Early contractor selection accompanied by a quicker start to, and completion of, the contract.
(2) The contractor's detailed pricing is known after the first stage, following the receipt of competitive offers, and this can be used in determining the rates at the second stage. Knowledge of the contractor's pricing methods is of considerable value when negotiating additional items in the contract.
(3) There are benefits at the design stage with the availability of the expertise and experience of the contractor and his organisation.
(4) Construction may start before the design is complete, although there are dangers inherent in this procedure.[6]

SELECTION OF TENDERERS

Pre-qualification of Contractors

With selective tendering, pre-qualification of contractors is normally a prerequisite to assist in compiling a list of firms qualified to receive invitations to tender. Contractors invited to pre-qualify are asked to submit details of their experience relevant to the specific project under consideration. The amount of information required should reflect the technical content and scope of the works in question. The factors to be considered can be categorised under three broad heads:

(1) *The contractor's financial standing.* It is important to know whether the firm is financially stable and/or has the guaranteed backing of a larger group to withstand any possible financial problems that may occur during the contract. The investigation normally includes examination of annual reports, in the case of a public company, and a letter or confidential report from the firm's bank. However, the balance sheet or financial report of a construction company needs careful scrutiny by a person familiar with the construction industry as, for example, the existence of a large order book may not necessarily be a desirable feature.[1]

(2) *Technical and organisational ability.* It is necessary to establish whether the firm has adequate capacity and ability and a satisfactory management structure to undertake the works at the appropriate time. Enquiries made of other employers and engineers can assist in assessing this aspect based on past experience.

(3) *General experience and performance record.* The engineer will wish to determine whether the firm has had sufficient experience in the particular type, scope and size of project to be undertaken, and has a satisfactory performance record. This knowledge is best obtained by interviewing the contractor rather than relying entirely on published documents or the views of others.[3]

A CIOB paper[1] recommends that contractors be asked to attend meetings relating to the project, to meet the design team and, on major works, be interviewed by the engineer and members of the design team and the employer, in order to satisfy them as to their ability to carry out the project effectively. Employers ideally should meet the individuals that will be responsible for the project, and other key people in the contractor's team.

The extent to which contractors should be asked to pre-qualify will depend on the nature of the work and the information that is already available to the engineer and/or employer. Contractors whose qualifications and past performance records are already well known should not be required to pre-qualify on every occasion.

Haswell and de Silva[7] have listed the type of information that prospective tenderers might be required to supply, and the principal items are as follows:

(1) legal status of the company;
(2) particulars of holding, subsidiary or associated companies;
(3) the latest published audited annual report and accounts of the company and those of the previous two or three years;
(4) annual turnover of the company relating to works of civil engineering construction;
(5) statement of construction projects of a similar nature to the proposed scheme executed by the company with their respective values;
(6) number of persons employed in the company's head office and branch offices, excluding those employed in site offices;
(7) names of directors;

  (8) numbers employed in management and administrative positions;
  (9) numbers employed in design and drawing offices of the company;
 (10) construction plant and labour resources of the company; and
 (11) references, membership of trade associations, permission to visit some of
       the works executed by the company, and other relevant data.

Horgan[8] has described how generalised references concerning contractors can
be of only limited value. Letters seeking references should request information
on specific aspects of a contractor's performance and could take the form of a
questionnaire. Matters which particularly need clarifying are, for example, the
contractor's record in respect of the following aspects:

  (1) management and administrative efficiency;
  (2) ability to control and organise contracts and efficiently integrate labour
       resources;
  (3) extent and scope of claims;
  (4) extent and dependence on sub-contractors;
  (5) ability to work accurately to drawings and extent of abortive work;
  (6) ability to meet target dates, show a sense of urgency and avoid excessive
       waiting time;
  (7) ability to co-ordinate the use of diverse earth-moving plant;
  (8) history of strikes and labour unrest;
  (9) details of any special or unusual items of plant which can be provided;
 (10) performance, efficiency and reliability of the plant supplied;
 (11) availability of spare parts and effectiveness of maintenance services in rela-
       tion to plant;
 (12) extent of collaboration and liaison with employer and engineer; and
 (13) any technical aspects where the contractor's competence may be in doubt.

Approved Lists

The policy of maintaining standard lists of approved contractors, as opposed to
compiling selected lists on an *ad hoc* basis for each project, is influenced mainly
by the size, nature and continuity of the employer's programme of works. The
principal aim is to identify those firms who have the necessary technical and
financial resources to complete the contract satisfactorily, often using the pre-
qualification process described earlier. The engineer should take full account of
the views of the employer, who may have firm ideas as to contractors he favours
and those he wishes to avoid.[8]

Standing lists are normally stratified by types and categories of work and
ranges of value. An ICE publication[3] emphasises the importance of preventing
the decline of standing lists by failing to provide an incentive for those included
on the list to maintain the necessary standard. Individual performance should be
reviewed at the end of each contract and the opportunity taken to transfer the

firm to another category, or remove it altogether as appropriate. The applications by new firms for inclusion should be considered at least annually.

Number of Tenderers

The main objective of selective tendering is to limit the number of contractors tendering to a realistic level. It has become generally accepted practice with civil engineering projects to invite from four to eight contractors to tender for a project.[3]

The number of contractors invited to tender is influenced by the size of the project. Tendering is an expensive activity and the costs rise with increases in the size and complexity of projects. Hence the larger the project the fewer should be the number of tenders invited, as only one contractor will be successful and thus recover his tendering costs from the particular project. There is, however, the danger that if too few *bona fide* tenders are received in response to an enquiry, the employer and the engineer face the additional expense of repeating the tender arrangements, accompanied by a loss of planned project time. The length of the list of selected tenderers is therefore a compromise between a satisfactory spread of offers and a minimum number of out-of-pocket contractors.[8] Full and fair competition should be provided in the interests of the employer and the construction industry and, where the employer's programme of works encompasses a large number of contracts, the system of selection should ensure that equal opportunities are offered to all qualifying contractors.[3]

It is interesting to note that a NJCC Code[4] recommended that the maximum number of tenderers from selected lists for building projects should be as follows: contracts up to £50 000: 5; between £50 000 and £250 000: 6; between £250 000 and £1m: 8; and £1m plus: 6. A caveat is included regarding specialised contracts in engineering services which entail relatively high tendering costs, where the maximum number of tenderers for contract values above £250 000 should be reduced to six and four respectively. Once the list is finalised, one or two further names should be appended so that these may replace any firms on the list who do not take up the preliminary invitation to tender.

Preliminary Enquiries

With selected lists of contractors being relatively short, it is customary to make preliminary enquiries as to a firm's intention to tender for a specific contract. The purpose of the enquiry is to ensure that the required number of completed tenders are received and, in the event of a firm declining, that another from the selected list can be substituted to make up the desired number. A contractor's acceptance of the tender invitation is often dependent on his workload at the time, and it is far better that a busy contractor should decline the offer to tender than to submit an unrealistically high cover price. Furthermore such action should not prejudice the firm's opportunity of tendering on future occasions and this should be made clear in the letter of enquiry.

The enquiry letter should always contain adequate information to enable the contractor to fully appreciate the nature and scope of the work and the expected period during which the work will be carried out. He will then be able to fully consider the project in relation to his available resources. The following items indicate the type of information that may be supplied to prospective tenderers.

(1) employer's name and title of the project;
(2) location of the site;
(3) outline particulars of the contract works, including any abnormal or unusual features;
(4) extent of any design, development or drawing work to be carried out by the contractor;
(5) an indication of the extent of the works, such as principal quantities and/or overall dimensions of the major component parts;
(6) expected date of issue of tender documents;
(7) intended date for return of tenders; and
(8) expected contract programme, date of placing contract, commencement date and completion date.[8]

SUPPLY OF INSTRUCTIONS AND OTHER INFORMATION TO TENDERERS

Instructions to Tenderers

The instructions to tenderers are normally issued with the tender documents for the purpose of drawing attention to the conditions applying to the invitation to tender and the procedures to be followed or complied with by the contractor when compiling and submitting his tender. They are provided solely for the purpose of conveying information and instructions which apply during the tender or pre-contract period. They usually stress the importance of all entries and signatures by the contractor being clear and unambiguous, and that any alterations must be initialled.

An ICE publication[3] lists the following matters to which the instructions will normally draw attention.

(1) Inclusion or otherwise of a contract price fluctuation clause.
(2) List of forms to be completed and submitted with tenders.
(3) Insurance requirements and procedures.
(4) Method of dealing with queries.
(5) Acceptability or otherwise of qualified tenders.
(6) Acceptability or otherwise of alternative proposals.

(7) Method of dealing with errors.
(8) Procedure for publishing the tender results.
(9) Requirements for presenting rates and prices in the bill of quantities, such as the use of decimal places.
(10) Any matters that may influence the employer's decision to enter into a contract.
(11) Procedures to be adopted in presenting and submitting tenders.
(12) Venue, time and date for return of tenders and applications for an extension of time for tendering.
(13) Approximate commencement date envisaged and completion times.
(14) Arrangements for inspecting the site during the tender period.

Site Information

All factual information relating to the site and ground conditions obtained by the engineer should be included in the tender documents or otherwise made available for inspection by tenderers. This will include such information as borehole logs, rainfall records, river levels, tidal records, earthquake information and any special requirements or restrictions that may need to be considered.

Supply of Documents

The number of sets of tender documents supplied to tenderers is normally not less than two, and one copy of the priced bill of quantities is often required to be returned with the tender.

Drawings, including bar bending schedules where available, should always be issued with the tender documents. Notes on drawings are added to amplify the design details; they should be legible and certainly not conflict with the specification.

ACTION DURING THE TENDER PERIOD

Tender Period

Tender periods must be sufficient to allow tenderers to familiarise themselves with the documents and sites, obtain quotations, decide the best constructional methods, price the general items, method related charges and unit rates, and insert a lump sum addition or deduction against the adjustment item in the grand summary in adjustment of the total of the bill of quantities. Hence, this period is one of intense activity by contractors and it is in the employer's best interests to allow adequate time from the outset. The length of the tender period is influenced by the value and complexity of the project and the extent of the negotiations to be undertaken by tendering contractors. It should be not less than four weeks and for large projects a period of eight weeks is recommended.[3]

The disadvantage of a short tender period is that the tenderers will have insufficient time to plan the work properly and economically and price it in a realistic manner. If time is short, tenderers are likely to qualify their tenders and any time saved may then be lost during the tender adjudication period and in delays which may occur during the construction period as a result of inadequate pre-tender planning.[7]

Queries and Pre-tender Meetings

Tenderers should aim to clarify all doubtful points as soon as possible and before the submission of tenders. Where queries relate to possible ambiguities, the need for supplementary information or errors in the documents, any verbal replies by the engineer should be recorded and then confirmed in writing by means of addenda notices or amendments issued to all tenderers.[3]

Pre-tender meetings are sometimes arranged to clarify points of doubt and uncertainty arising during the tender period. They are best arranged in the form of a group meeting of all tenderers, possibly on site, and all information provided on these occasions should be properly recorded and confirmed in writing to all tenderers. Where this information has contractual significance it needs to be incorporated in the contract documentation.[3]

Amendments

Amendments to the tender documents should be issued only if the engineer is convinced that the pricing of the tenderers will be so seriously affected as to distort the balance between the tenderers' offers. Unless significant changes are necessary, amendments are better delayed and dealt with either as post-tender clarification or as part of the administration of the contract after acceptance.[3]

Tenderers should be notified immediately if it is decided to issue any amendments to the contract documents. Where the amendments are significant, an extension to the tender period may be necessary. Amendments made to the documents are best prepared in the form of addenda letters or notices, which should be numbered consecutively and accompanied by an acknowledgement slip.[3]

Receipt of Tenders

Tenders should be sent by registered post, or recorded delivery service, or be delivered by hand in a plain sealed envelope, entitled as directed but not bearing the name of the tenderer. Tenders are confidential and should remain unopened and secure until the designated time for opening. Tenderers have the right to modify their tenders in writing at any stage before the prescribed opening time.

Tenders received after the prescribed date and time are invalid and should not be considered unless:

(1) a cable or telex stating the tender sum is received on time, and
(2) there is clear evidence, such as a postmark, that the complete tender documents have been despatched within a sufficient margin of time to presume its due arrival.

Completion Date

Varying opinions are held about the advisability of the time for completion being determined by the employer, as in the majority of circumstances it can be argued that contractors are better able to decide what is a reasonable completion time. Furthermore, time and cost are interrelated. One approach is to insert a completion date in the contract documents, but to inform contractors that they may submit an alternative tender based on their own selected contract period. However, the comparison by the engineer of tenders based on different contract periods is very difficult.[9]

ALTERNATIVE TENDERS

With civil engineering contracts it is not unusual for the tender documents to expressly permit the submission of alternative designs with their prices. Furthermore, as long ago as 1964, the Banwell Committee,[10] when considering the placing and management of both building and civil engineering contracts, expressed the view that 'if a firm has the initiative to produce a novel and possibly better technical solution, fully documented, to a problem, we cannot see why this should be ignored or disclosed to rival tenderers. Such alternatives should be considered on their merits.'

A contractor may wish to carry out the works in a different sequence and/or by different methods from those shown in the tender documents. Ideally, tenderers should not be discouraged from submitting alternatives, provided they also give a *bona fide* offer for the tender scheme, and it is clearly understood that the employer is free to accept or reject the alternative. Although they make the task of the engineer in selection more difficult, alternative tenders should as far as practicable be encouraged, as there may be a considerable advantage to the employer in the alternative designs, sequences and methods of construction offered by the tenderer, based on his practical experience of carrying out similar projects.[9]

The instructions to tenderers should state whether offers based on alternative designs will be considered as potential options and, if so, should advise the contractor to ascertain from the engineer details of any special design criteria and requirements that will affect the alternatives. Where alternative designs are admissible, certain procedural rules should be observed to ensure a fair and proper adjudication of the tenders. These rules, incorporated in the instructions to tenderers, could embrace the following matters as listed in an ICE publication on Tenders for Civil Engineering Contracts.[3]

(1) A tender based strictly on the tender documents is also submitted.
(2) A tenderer is to give at least two weeks notice before the date of return of tenders of his intention to submit an alternative.
(3) An alternative tender is to be accompanied by supporting information such as drawings, calculations and a priced bill of quantities addendum, in order that its technical acceptability, construction time and economics can be fully assessed.

SELECTING INVITATIONS TO TENDER

Contractors should accept invitations to tender only when they are satisfied that they have adequate time available in the estimating department to allow proper preparation. Overloading the estimating department can lead to errors of judgement and arithmetic. Estimating is an exacting and precise activity and the submission of unrealistic estimates can jeopardise the financial stability of a company.

Furthermore, the contractor submitting a tender must be satisfied that he has the necessary technical and financial resources to complete the project on programme. It is bad policy to submit tenders below cost when there is a severe shortage of work, in this way hoping to recoup the loss when prices improve, as this is very unlikely to happen.

DISTINCTION BETWEEN ESTIMATES AND TENDERS

The CIOB *Code of Estimating Practice*[11] distinguishes between estimating and tendering. *Estimating* is defined as the technical process of predicting costs of construction, while *tendering* is a separate and commercial function based upon the nett cost estimate, and culminates in an offer to carry out defined work under prescribed conditions for a stated sum of money.

Harrison[12] has aptly described how an estimate is a reasonably accurate calculation and assessment of the probable cost of carrying out defined work under known conditions. If the estimate is not reasonably accurate it will be of little value and could subsequently result in major problems for the contractor. However, the degree of accuracy obtainable will be affected by the particular circumstances of each project.

An estimate involves both calculation and assessment, and both technical data and human judgement of circumstances and probabilities must be brought together in its production. The wording 'probable cost' recognises that an estimate is a prediction of likely outcome, which must be based on a judgement of probabilities and risk. The terms 'defined work' and 'known conditions' indicate that before an estimate can be produced there must be adequate data supplied to form a basis for the calculations and assessments.

Having regard to the extent to which human judgement is inextricably bound up with the estimating process and the immense range of variables to which a construction project may be subject, it is understandable that there are occasionally wide variations between tenders for a project, or between the estimate and the actual cost.[12]

PRELIMINARY ESTIMATING PROCEDURES

Examination of Preliminaries and Project Information

As soon as the tender documents are received, a senior member of the contracting organisation, such as a director or the chief estimator, should examine the general items and preliminaries bill to ascertain whether there is any alteration to the prescribed form of contract or any unusual or unfair conditions. If these provisions were to be unduly onerous and to introduce exceptionally high risks, the contractor might decide not to submit a tender.

The estimator should extract details from the project information which:

(1) affect the contractor's intended method of working;
(2) impose restrictions of any kind;
(3) affect access to the site;
(4) interrupt the regular sequence of trades or construction activities;
(5) affect the duration of the project;
(6) require specialist skills, processes or materials;
(7) have a significant effect on the programme; or
(8) are of major cost significance.[11]

These items affect costs and will therefore be of concern to the estimator in the preparation of the estimate.

The estimator must liaise with, and co-ordinate the views of, the various members of the contractor's team who examine the project information. He must provide management with a realistic appraisal of this information to enable a decision to be made on whether or not to tender for the project.[11]

All tender documents must be examined in detail by the estimator. He has a responsibility to ensure that all other members of the contractor's organisation involved in purchasing, programming and other associated activities are provided with copies of all relevant project information.

The estimator will look for any factors which may influence his approach to the pricing of the project, such as:

(1) standard and completeness of drawn information;
(2) tolerances required;
(3) clarity of specification requirements and the quality required;
(4) buildability;
(5) extent of use of standard details and amount of repetitive work;
(6) amount of information concerning ground conditions and sub-structural work; and
(7) problem areas and constraints on construction imposed by the design.[11]

From his examination of the drawn information, the specification and measured items in the bill of quantities, the estimator will acquire an understanding of how the project is to be constructed and the quantity and quality of resources required. It is unfortunate that while designers frequently take many months and sometimes even years to formulate the design, the estimator and his team will be required to assimilate all the information produced, decide on how the project will be constructed and estimate its cost of construction, often within a tender period of a few weeks.

The estimator lists any outstanding information together with any queries raised by other members of the contractor's team who have examined the documents. It is good practice to channel all queries concerning tender information to the estimator for resolution. Ease of communication and good management practice dictate that only the estimator should deal with the engineer at this stage.[11]

The estimator must check that the items of work covered by prime cost (PC) sums are clearly stated and that the work can be identified within the project. Other items to be checked include:

(1) whether nominated sub-contractors are known;
(2) that nominated sub-contractors will be completing design indemnity warranties and that the contractor assumes no design responsibility for such works;
(3) the extent of progress of such design work and co-ordination with other consultants' drawings;
(4) that attendance is adequately identified;
(5) that adequate allowance has been made for the contractor's profit addition and that discounts are to be allowed;
(6) that nominated sub-contractors can conform with the requirements of the contractor's tender programme; and
(7) that contractor's work requirements associated with that of specialists are clearly defined and measured in the bill of quantities.[11]

Method Statement and Tender Programme

It is essential that an early meeting is held between the estimator and those responsible for the programming and construction of the project, to establish initial proposals relating to the method of construction. The following points are likely to be considered when deciding on the method of construction:

(1) site location and access;
(2) degree of repetition;
(3) space available for storage, hutting and movement on site;
(4) adjacent buildings and structures;
(5) company's experience of the type of project to be constructed;
(6) availability of labour;
(7) availability of materials;
(8) extent of specialised work and its relationship to the general construction;
(9) amount of work to be sub-contracted;
(10) extent to which the design dictates construction method, such as formwork striking times and special sequence of construction; and
(11) plant requirements.[11]

A CIOB publication on estimating[11] describes how decisions concerning the resources to be used on the project will take into account such factors as:

(1) location and availability of labour and management within the company;
(2) cost of recruiting additional labour, its availability, quality and quantity;
(3) amount of work to be sub-contracted;
(4) plant available within the company;
(5) availability of plant outside the company;
(6) availability of materials, including long delivery items;
(7) current and future projects in the area which may affect the supply of basic resources;
(8) quality of workmanship required;
(9) special requirements of the project, such as special plant or skills needed;
(10) overlap of operations needed to meet programme requirements;
(11) materials handling on site, storage, distribution and waste, time span of the project and seasonal influences on method of construction; and
(12) quality and complexity of the work.

Alternative operational methods, site organisation and work sequences should be evaluated at this stage and decisions made on the intended method of construction. Confirmation of these arrangements may have to await a visit to the engineer and/or the site visit.[11]

The contractor prepares a tender programme and a method statement. If the tender is successful the programme represents the contractor's intentions at the time of tender and upon which the pricing of the works were based.

The method statement outlines the sequence and methods of construction upon which the estimate is based. It will show how the major elements of work will be handled and will indicate areas where new or different methods will be used. It should be supported with details of cost data, gang sizes, plant and supervision provisions.[11]

The main purposes of the method statement are:

(1) to establish the principles on which the estimate is based; and
(2) to acquaint construction personnel of the resource limits which have been allowed in the estimate and to describe the method of working envisaged at the tender stage.[11]

Site Visit

Following the preliminary assessment of the project, the estimator will visit the site often accompanied by other members of the contractor's team. The party will normally extend their visit to embrace the general locality to examine works in progress, particularly earthworks, and other matters of significance.

The site report will include details of the site and its surroundings, access, topographical features, including trees and site clearance work, ground conditions, groundwater level if it can be determined, existing services, including any overhead cables, possible security problems, facilities for disposal of surplus spoil, labour situation, availability of plant and materials, weather conditions and effects of bad weather, any special problems such as tides, high winds, flooding, noise control and road diversions, any constraints on working on site, such as restrictions on the area available for tower cranes and other plant, and space for site huts, storage compounds and the like.[13]

Labour Availability

The bill cannot be priced until the local labour situation has been investigated. Where the work is to be carried out in an area well known to the contractor, he should have sufficient information in his records. Where the site of the project is in a location with which he is unfamiliar, then investigations will be necessary, usually starting with the local Job Centre. The estimator should supply details of the approximate numbers of operatives required in each trade.

Plant Selection

The early selection of the most suitable plant is necessary, once the method of construction has been decided. This will normally entail discussions between the estimator, contracts management and, if possible, the person who will be the company's site representative. The choice of plant will also be influenced by the availability or otherwise of suitable items of plant in the company's ownership, and a careful study of the site. The decision reached at the meeting will be recorded and acted upon by the estimator in preparing the estimate.

PREPARATION OF ESTIMATES

Components of an Estimate

An estimate for a civil engineering project comprises a combination of diverse components which are now listed.

(1) directly employed labour values;
(2) value of materials to be purchased direct;
(3) plant values;
(4) value of works to be placed by the contractor with sub-contractors of his own choice;
(5) value of supplies to be obtained from sources to be nominated by the engineer;
(6) value of works to be carried out by sub-contractors nominated by the engineer;
(7) value of provisional sums;
(8) value of prime cost sums;
(9) value of preliminaries and general items in the bill of quantities; and
(10) overheads and profit.

Estimating Procedure

Once the decision has been made to tender for a project, the contractor's estimator assembles all the necessary information by inviting quotations from suppliers and sub-contractors, examining the drawings, visiting the site and discussing with management the construction methods to be used, as described earlier in the chapter. He will then determine the all-in hourly rates for labour and plant, analyse the quotations for materials and sub-contractors' work and build up unit rates so that all the sectional work items in the bill of quantities can be priced. At this stage, a pre-tender construction programme may have been agreed or, alternatively, it may be left in abeyance until the main sections of the bill are priced and the estimator is more familiar with the construction details of the project. The estimator then usually follows by estimating the cost of overheads and pricing the preliminaries and general items in the bill, prior to completing the summary to the bill and the adjudication of the estimate by management to arrive at the tender total. The procedure outlined is not used universally but provides a sound plan of action.[14]

In pursuance of good estimating practice, all necessary information is obtained, alternative methods of constructing and pricing the project are considered, the most appropriate method adopted in each case, accurate and careful appraisal and evaluation is carried out at all stages of the preparation of the cost estimate for the selected method, and the whole process is checked at adjudication prior to deciding upon the tender offer. Despite all the care and thought given to estimating and tendering, contracts often lose money, not because of inefficient contract management or poor productivity, but because the cost estimate does not properly reflect the true cost.[14]

It is suggested by Sharp[14] that the tender total may be incorrect because various aspects of the estimating procedures are not properly related or reconciled. Some examples will serve to illustrate his reasoning.

(1) All-in hourly rates for craft operatives often include provision for overtime working partly to ensure completion of the project within the contract period but, on occasions, more importantly to provide an additional incentive to attract and keep the operatives on the project. Subsequent calculations of unit rates for items such as brickwork, using gangs of craft operatives and labourers in varying ratios, may not include for labourers working overtime to coincide with the bricklaying operations. Furthermore, the inclusion of an allowance for inclement weather should be assessed according to the timing of the contract, otherwise the cost estimate will be too high for work undertaken in the summer and too low for that in the winter.

(2) the outputs forming the basis of unit rates are often extracted from schedules prepared by the estimator based on his personal experience, or from data contained in the firm's costing system or a combination of both. Having arrived at these unit rates, they are not always adjusted to suit the varying quantities of work to which they are applied.

(3) The prices inserted for project overheads generally consist of a combination of lump sums and costs that are time-related. It is essential to assess accurately the contract construction period and to determine the sequential phasing of the various sections of work, so that such items as site supervisory costs and scaffolding can be realistically priced. In arriving at the contract construction period, some consideration will be given to the fact that overtime working has been allowed for in the all-in hourly rates for operatives and that a specific level of output has been decided upon in building up unit rates, but, as Sharp[14] has described, seldom are these major components reconciled correctly. This highlights the need for detailed planning of the project to be carried out in parallel with the preparation of the estimate in order to effectively reconcile the two functions and thereby produce a realistic cost estimate.

The CIOB Code of Estimating Practice[11] describes how the estimator must have full responsibility for managing the production of the estimate. The initiation and control of enquiries, quotations and programme are important aspects of the estimating process. The estimator must co-ordinate the design of temporary works and liaise with general management on construction methods and available resources.

The Code[11] also describes how an estimate must be prepared in a way that is explicit and consistent, and which takes into account the methods of construction and all circumstances which may affect the execution of the work on the project. A realistic estimate can be obtained only when each activity is analysed into its simplest elements and the cost estimated methodically on the basis of factual information. Another important aspect of the estimator's work is to identify areas of risk and matters which may affect significantly the costs of a project, for consideration by management at the adjudication stage.

Preliminaries and General Items

It is customary for the first section of a civil engineering bill of quantities to encompass preliminaries and general items. These items can be classified under a number of broad heads, such as:

(1) contractor's site oncosts — time-related and non time-related;
(2) employer's and engineer's site requirements — time-related and non time-related;
(3) other services, charges and fees;
(4) temporary works other than those included in unit costs; and
(5) general purpose plant other than that included in unit costs.

The estimator can price these items or, alternatively, leave them unpriced and cover the costs elsewhere, such as in the unit rates. The time-related items are best calculated on the basis of the time for which the service is required, such as site staff salaries, plant maintenance, rental charges for site huts, provision of small tools and equipment for general use on site, and hire and operation of traffic lights. Non time-related items encompass such matters as the erection and dismantling of site huts, installation of site water supply including connection charges, progress photographs, and clearing of the site on completion.[15]

Labour costs

The cost of labour will be the recognised basic rates for skilled and unskilled operatives, plus the appropriate allowances prescribed by the Working Rule Agreement for the Civil Engineering Industry. In addition, all statutory payments to be made by the employer should be calculated, plus those under any voluntary or trade agreement. The aim must be to arrive at a labour rate per hour which is realistic and which reflects the actual cost of labour to the contractor including plus rates and guaranteed bonus. The following elements are normally included in the inclusive or all-in hourly rate.

(1) basic wage rates and guaranteed bonus;
(2) travelling time, fares and subsistence allowances (sometimes contained in project overheads as it is a variable item);
(3) holidays with pay scheme covering public holidays and annual holidays;
(4) tool allowance;
(5) sick pay;
(6) employer's National Health Insurance contributions;
(7) employer's liability and third party insurances;
(8) Construction Industry Training Board (CITB) levy.[15]

The additional labour payments amount to a sum in excess of 100 per cent of the basic hourly wage rate, and, as shown in the preceding list, include costs associated with the Working Rule Agreement and certain overhead costs incurred by the employer. Other costs, which will be variable and may be specific to a project, and some time-related costs are normally included in project overheads. These additional costs include such items as guaranteed time, pensions, possibly daily travel allowances and fares, lodging/subsistence allowances, abnormal overtime to meet time targets, attraction money and special severance payments.[11]

It may not be possible to accurately assess all the labour oncosts early in the estimating process, particularly as the total labour requirements will not have been quantified. There are distinct benefits in omitting the more variable items from the all-in labour rates and to include them later in the project overheads. It is often a matter of opinion and company preference where many of the items are priced and there cannot be a universal practice. The main consideration is to make realistic assessments of all the costs involved and to incorporate them in the estimate.[11]

Materials Costs

The CIOB Code of Estimating Practice[11] has highlighted how the contractor's success in obtaining a contract can depend upon the competitiveness of the quotations obtained for materials, plant and sub-contract items. The responsibility for this function varies from one organisation to another. Often the estimator prepares the enquiry documents, selects the firms to receive enquiries and reconciles the quotations received. In other cases, some, or all of these functions, will be performed by the buying department, who will then supply the estimator with a selection of fully reconciled quotations. The quotations should be recorded methodically in an appropriate quotations register.

Enquiries to suppliers of materials should contain certain information, of which the following comprise the major items.

(1) title and location of the work;
(2) specification, class and quality of the material;
(3) quantity of the material required;
(4) likely delivery programme and any special delivery requirements;
(5) access to site and any restrictions;
(6) date by which quotation is required;
(7) period for which quotation is to remain open;
(8) whether a fluctuating or firm price is required;
(9) discounts required; and
(10) person in the contractor's organisation to be contacted when queries arise.[11]

Before selecting the quotations of materials for use in pricing unit rates, they should be checked to ensure that they meet the following criteria:

(1) the materials comply with the specification;
(2) the materials will be available in sufficient quantities to meet the requirements of the construction programme;
(3) no special delivery conditions have been imposed by the supplier;
(4) the method and rate of delivery complies with the contractor's requirements;
(5) the conditions contained in the quotation do not constitute a counter offer which is at variance with the terms and conditions of the enquiry;
(6) the quotation is valid for the required period;
(7) prices are given for small quantities where applicable;
(8) discounts conform to the requirements of the enquiry; and
(9) requirements concerning fixed or fluctuating prices are satisfied.[11]

When building up the unit rates, material prices should include the basic price, less discounts retained by the contractor, allowance for waste, unloading, stacking, storing, distributing around the site and the return of crates or packings where appropriate.[16]

Plant Costs

The contractor's plant requirements will be derived from the method statement and the programme. These documents will establish the basic performance requirements of the plant and in many instances will identify the specific items of plant needed for the works. The period for which the plant is required on the site can be obtained from the tender programme. The estimator normally starts by preparing a schedule of plant requirements, listing the type, performance and time requirements. The plant can be conveniently categorised into:

(1) mechanical plant with operator;
(2) mechanical plant without operator; and
(3) non-mechanical plant.[11]

A note should be inserted on the plant schedule of any additional requirements associated with a particular item of plant, such as power supply for a
tower crane or temporary access roads for erection purposes. The contractor has
three main choices available to meet his plant requirements:

(1) purchasing plant for the contract;
(2) hiring existing company owned plant; and
(3) hiring plant from external sources.[11]

Plant owned by a contractor can be classified under two broad heads:

(1) Small plant and tools which are the subject of a direct charge to contracts
     and, for estimating purposes, are normally allowed for as a percentage of the
     labour cost in site oncosts.
(2) Power driven plant and major items of non-mechanical plant such as steel
     trestles, scaffolding and gantries. Such plant is normally charged to the con-
     tract on a rental basis, except in the case of plant specially made or purchased
     for a specific operation. The latter plant is normally charged in full to con-
     tracts and allowance made for disposal on completion, often at scrap value.[15]

A wide range of plant is readily available from plant hire companies. It is not
usually economical for contractors to own plant unless they can ensure 75 to 80
per cent use during the contractors' normal working hours. Where a contractor
owns plant, it is important that he maintains reasonably accurate records of the
working hours and costs of upkeep, in order that realistic charges can be determined for each item of plant. Table 3.1 shows a method of calculating the rental
charges for contractor owned plant.[15]

The CIOB Code of Estimating Practice[11] describes how quotations for plant
should be carefully checked to ensure that the plant meets the contractor's
requirements, and the following matters need to be clarified.

(1) the plant complies with the specification;
(2) it is available to meet the needs of the construction programme;
(3) delivery and collection charges can be identified;
(4) where appropriate, all operator costs are included and that operators will
     conform to the intended working hours of the site;
(5) any attendance or supplies to be provided by the contractor are clearly
     identified;
(6) maintenance responsibilities, charges and liabilities are identified;
(7) the quotation is valid for the required period;
(8) the quotation conforms to the terms and conditions of the enquiry and does
     not represent a counter offer; and
(9) requirements concerning fixed or fluctuating prices are met.

Table 3.1   Rental charge for a crawler hydraulic shovel loader

	£
Purchase price	42 000.00
Resale value after five years	10 000.00
Loss in value	£32 000.00
Use in hours per year	1700
Total hours used over five years	8500
Hourly cost	£

$$\frac{\text{depreciation}}{\text{total hours}} = \frac{£32\ 000}{8500} \qquad\qquad 3.76$$

$$\frac{\text{maintenance}}{\text{total hours}} = \frac{£3700 \times 5}{8500} = \frac{£18\ 500}{8500} \qquad\qquad 2.18$$

Finance, assuming 14 per cent per annum interest charges

$$\frac{£32\ 000}{2} \times \frac{14}{100} \times \frac{5}{8500} \qquad\qquad 1.32$$

Licences, insurance, etc.
$$\frac{£750 \times 5}{8500} \qquad\qquad 0.44$$

Total cost per hour	7.70
Add for head office and administration charges (20 per cent)	1.54
Rental cost per hour	£9.24

Sub-contractors

It is quite common for the value of directly employed and nominated sub-contractors' work to form a significant part of the total estimate. The firms quoting are usually requested to insert their rates in copies of the relevant sections of the bill of quantities which are sent to them. On receipt the sub-contractors' prices will be checked and any special conditions in the sub-contract noted which may affect the contractor and involve him in additional expense. Among such conditions might be the provision of special scaffolding or means of access, lighting, heating or electrical power, hoisting facilities, storage and protection.[16]

Prime Cost and Provisional Sums

In the case of prime cost and provisional sums, the elementary precaution should be taken of checking to ensure that the sums of money included in the text are extended into the pricing column. Following each prime cost item there will be provision for the addition of profit and attendance. Profit is normally calculated on a percentage basis while attendance will be assessed on the cost of the services to be provided and entered as a lump sum.

Unit Rates

In calculating unit rates for insertion in the bill of quantities, it is essential that careful consideration is given to every factor which may influence the cost of the work. As stated in the CIOB Code of Estimating Practice,[11] there can be no substitute for comprehensive company data and feedback from previous work of a similar nature to the project being priced. Unit rates for measured items in the bill can consist of any, or a combination of, the basic elements of labour, materials, plant, sub-contractors' items, overheads and profit. It is recommended that each element is analysed and estimated separately so that the total cost of each element can be considered by management.

The cost or output on previous projects depends upon many variables, and attention must be paid to the conditions prevailing on each project and the levels of incentives operated to achieve the particular standard. These conditions must be compared with those expected to be encountered on the project under consideration. Any differences between estimated and actual costs or outputs on previous projects must be analysed and the causes noted. The estimating data must be subject to regular updating.[11]

The term 'labour constant' is something of a misnomer in that labour, like many other factors in estimating, is anything but constant. A contractor can only price on what he believes his organisation, site supervision and quality of labour can produce, based upon past performance suitably adjusted for the new situation.

Spon's Civil Engineering Price Book[15] contains numerous items of unit rates which provide useful guidelines but require adjusting for local conditions, updating and possibly the inclusion of head office overheads and profit to provide a check on the contractor's unit rates. They can, however, be useful in the preparation of preliminary estimates if used with care and suitably adjusted. The prices of materials and labour rates are fluctuating continually.

A relatively straightforward price build-up item is illustrated in table 3.2 to show a common method of approach.

**Table 3.2 Cost of one-brick wall in engineering bricks (class B)
in cement mortar (1:3)/m²**

	£
120 engineering bricks @ £210/thousand	25.20
Add waste 5 per cent	1.26
Cement mortar 0.08m³ @ £45/m³	3.60
Unloading bricks: labourer: 0.10 hrs @ £3.20	0.32
Bricklayer: 2 hours @ £3.70	7.40
Labourer: 1 hour @ £3.20	3.20
	40.98
Add head office oncosts and profit (15 per cent)	6.15
Cost of brickwork/m²	£47.13

Overheads and Profit

Harrison[12] has defined overheads as 'those costs incurred in the operation of a business which are not directly related to individual items of production'. There are two main groups of overheads:

(1) site overheads which include such costs as site supervisory staff, site build-ings, temporary roads and services; and
(2) head office overheads which cover the costs incurred in operating the busi-ness in its entirety and cannot be related directly to an individual contract, and include head office staff and buildings.

With site overheads each contract will have a calculated allowance in the tender, but head office overheads require a different method of assessment. One approach is to calculate the budgeted overheads as a percentage of budgeted turnover and to apply this to all contracts.

Farrow[16] has described how most contractors use systems which record past and present overheads, the projection of overheads for the future, and rate at which overheads are being recovered. An important part of a contractor's general overheads is the cost of financing construction works in advance of payment which needs to be calculated and included in the tender. The appropriate addi-tion for head office overheads varies with the extent of centrally provided services and the size of organisation, but could be in the range of 4 to 8 per cent of turnover.[15]

The amount of profit required by individual firms will also vary with the volume of work in hand, orders anticipated and the general tendering climate. It is customary to include an addition of 3 to 5 per cent of nett turnover.

Use of Computers in Estimating

McCaffer and Baldwin[13] have described how the use of computers in estimating work has developed with the availability of smaller and cheaper computers, and interactive programs, whereby the estimator can key in his instructions and receive a response on the visual display unit (VDU). The application of computer aided estimating offers the facility to retain detailed, meaningful and accessible records of the estimator's calculations and assembled data, which can also help with subsequent project control activities.

It is considered that a computer aided estimating system must be able to:

(1) calculate bill item prices from input data;
(2) apply calculated rates to all relevant bill items;
(3) provide an extension and summation of bill item rates to produce direct cost totals;
(4) provide a variety of report and bill listings for the estimator and other personnel;
(5) store data on different resources and their requirements for different constructional methods;
(6) store lists of all-in rates and materials and sub-contractors' prices;
(7) store the full build-up of each bill item with the facility to retrieve, check and recalculate the item if required;
(8) assist the estimator in communicating with other parties;
(9) maintain and extend the estimator's skill and knowledge of the construction processes; and
(10) restrict potential errors within the estimating process.[13]

COMPILING TENDERS

Tendering is the process of converting an estimate into a commercial bid. For this purpose it is necessary to assess competition in relation to previous tenders, with the aim of achieving a tender total low enough to beat competitors, yet high enough to secure a worthwhile profit. However, departures from strict commercial considerations are evident during depressed periods, with tender levels barely increasing despite spiralling cost increases of labour, materials and plant.[17]

Tassie[18] describes converting a nett estimate into a tender as the act of tender adjudication by senior management and directorate. He rightly believes that tender adjudication is concerned with obtaining the contract at the best price, securing it in competition by the smallest possible margin and on the best commercial terms. To achieve these aims necessitates a thorough appraisal of past performance of the company and tenders submitted compared with competitors, an assessment of the risks involved and the current and anticipated future workload of the firm.

The CIOB Code of Estimating Practice[11] envisages a two-tiered adjudication meeting in the larger organisations. The first stage is an estimate review by management and directorate and the second stage involves commercial decisions and a possible 'mark-up' by the directorate. The Code lists many matters which may be considered at this juncture. Variations can be made to unit rates, to the preliminaries and general items, as a percentage on the final summary page of the bill of quantities, or as a combination of these methods.

APPRAISAL OF TENDERS

Action on Receipt of Tenders

Tenders must not be opened before the prescribed day and time. It is advisable that the employer or his representative(s) should be present in addition to the engineer when the tenders are opened. It is good policy to tabulate the tenders, starting with the lowest, and listing the tender totals and the contract periods where these are not stipulated in the tender documents. If a tender is qualified in any way, this should be recorded. The tabulated list should then be signed by at least two persons present at the meeting.[6]

Where a tender is received after the stipulated date and time, it is generally advisable to reject it since there is a possibility that the tenderer may be aware of the amounts of some of the other tenders before submitting his own. The matter should be reported to the employer who will make the final decision. The NJCC recommendation is to promptly return the tender unopened to the sender.[4]

The notification of results cannot coincide with the receipt of tenders, as the tender documents, including the priced bills of quantities, have to be carefully scrutinised. It is good practice to request the three lowest tenderers to keep their offers open until the scrutiny has taken place.[9]

Scrutiny of Tenders

The examination of the priced bills by the engineer or his quantity surveyor is mainly a search for obvious errors. Inconsistent or unusual pricing methods may have a considerable influence on the sum finally paid to the successful contractor. For example, contractors occasionally inflate the rates for items of work to be carried out early in the contract and reduce the rates for later items. In this way the contractor will obtain higher payments than those to which he is really entitled during the early part of the contract and so improve his cash flow position at the employer's expense, while keeping his tender total the same.[6]

All the entries in a priced bill of quantities must be checked arithmetically as errors can occur in a variety of ways. Billed items may unintentionally have been left unpriced, there may be errors in item extensions (multiplication of quantities by unit rates), page totals, transfers of page totals to collections or summaries, in the Grand Summary or even in the transfer from there to the Form of Tender. Other pricing errors include the insertion of a cubic rate against a linear item and the inclusion of different rates for identical items in separate sections of the bill, without any apparent reason. An example of a genuine pricing error would be where provision of concrete items with a strength of 45.0 N/mm^2 and aggregate size of 20 mm are priced at £49.50/m^3 in twelve instances and £4.95 in four other places.[6]

Haswell and de Silva[7] have recommended the use of a schedule listing all items in the bill of quantities against which the rates and prices quoted by each tenderer are entered alongside the engineer's own rates and prices for these items. Ideally, the sequence should be the engineer's entries in the first pricing column followed by each of the tenderers in ascending order. A study of these entries will show whether any unit rate quoted by a tenderer is inconsistent or excessively weighted. A comparison of the totals tendered for each of the major components of the project will also be helpful in identifying the pricing patterns of the various tenderers.

Other factors requiring examination by the engineer include any qualification imposed by tenderers, programme requirements, deviations from the specification, mobilisation periods for staff and equipment, names of proposed subcontractors and major items of plant and equipment and their adequacy.

Correction of Errors

The ICE publication giving guidance on tenders for civil engineering contracts[3] states that 'If the engineer considers that a genuine mistake has been made, he should proceed to negotiate a correction only if he is satisfied that parity of tendering would not be breached, having regard to the fact that other errors, of which he is unaware, could be present in every tender. It requires careful judgement by the engineer to decide whether he should seek to change a tendered rate in one tender because of an error he deems to be genuine, whilst there is a probability that equally genuine errors remain undetected in other tenders, as to do so tends to undermine the whole basis of competitive tendering'.

The author favours the use of one of the two alternative methods of dealing with errors recommended by the NJCC,[4] of which the first is the most popular in practice.

(1) The tenderer should be given details of the errors and afforded an opportunity to confirm or withdraw his offer. If the tenderer withdraws, the priced bill of the second lowest tenderer should be examined and, if necessary, this tenderer will be given a similar opportunity. Where the tenderer

confirms his offer, an endorsement should be added to the priced bill indicating that all rates or prices (excluding preliminary and general items, contingencies, prime cost and provisional sums) inserted by the tenderer are to be considered as reduced or increased in the same proportion as the corrected total of priced items exceeds or falls short of such items. This endorsement should be signed by both parties to the contract.

(2) The tenderer should be given an opportunity of confirming his offer or of amending it to correct genuine errors. Should he choose to amend his offer and the revised tender is no longer the lowest, the offer of the firm which now has the lowest tender should be examined. If the tenderer elects not to amend his offer, an endorsement will be required as in method (1). If the tenderer amends his tender figure, and possibly certain of the rates in his bill, he should either be allowed access to his original tender to insert the correct details and to initial them, or be required to confirm all the alterations in a letter. If in the latter case his revised total is eventually accepted, the letter should be conjoined with the acceptance and the amended tender figure and the rates in it substituted for those in the original tender.

For example, if in a roadworks contract, the errors in total amounted to a reduction of £17 600, and the total of the measured works in earthworks, concrete roadworks and bridges amounted to £1 283 400. Then the percentage reduction to be applied to all the measured rates would be (17 600/1 283 400) x 100 = 1.371 per cent.

Tender Adjudication Report

Haswell and de Silva[7] have described how the engineer after examining and evaluating the tenders normally submits an adjudication report to the employer. This contains the engineer's findings on the tenders and may run into several volumes on very large contracts. The preparation of this report must be carried out well within the normal stipulated period of 90 days from receiving tenders to the award of a contract. The operative period is often six to eight weeks in order to allow a further six to eight weeks for the employer to complete his deliberations and award a contract.

A customary format for an adjudication report is:

(1) introduction;
(2) details of tenders received;
(3) initial elimination of tenders;
(4) comparison of tenders; and
(5) conclusions and recommendations based on technical and financial assessments of all tenders.

The adjudication report usually incorporates a series of appendices in the form of tables showing the details of the tender figures, tender totals, evaluated comparable tender totals, records of minutes of meetings, copies of relevant correspondence with tenderers after the receipt of tenders and other appropriate data.

NOTIFICATION AND ACCEPTANCE OF TENDERS

Notification of Final Results

An ICE set of guidance procedures[3] prescribes a useful acceptance framework, commencing with the notification of the successful tenderer as soon as it is decided to accept his offer. Should the employer decide not to proceed with the project, then all tenderers should be informed as soon as is practicable.

On the letting of the contract, the unsuccessful tenderers should be individually notified of the results, including the amount of the successful tender. All tenderers, including the successful one, should be supplied with a list of the names of all tenderers in alphabetical order, and a separate list of tender totals in ascending order of magnitude, thereby enabling individual tenderers to determine their ranking without specifically identifying other tenderers' offers. It is not, however, customary to circulate the names and amounts if the number of tenders received is less than four. It is good practice to refrain from issuing press notices until unsuccessful tenderers have been notified.

When writing to unsuccessful tenderers, it is customary, mainly for reasons of copyright, to request the return of all drawings and unused documents which were not submitted with their tenders.

Letter of Intent

Where it is not immediately possible to issue a formal letter of acceptance to the successful tenderer, it is good practice to send a letter of intent to enter into a contract. The letter is often issued with the main objective of authorising the contractor to carry out certain preparatory work before the formal acceptance of the tender and this will render the employer liable for any financial consequences that follow. The letter should therefore be clear as to its intentions and should ideally contain the following:

(1) a statement of intent to accept the tender at some future date;
(2) instructions to proceed, or not to proceed, with the ordering of materials, letting of sub-contracts and the like;
(3) a statement that if the tender is subsequently not accepted, or the letter of intent is withdrawn, then the costs legitimately incurred by the contractor will be paid by the employer;

(4) a limit to financial liability before formal acceptance of the tender;

(5) a statement that when the tender is formally accepted by the letter of acceptance, this will render the letter of intent void; and

(6) a request to the contractor to acknowledge receipt of the letter of intent and to confirm his acceptance of its conditions.[3]

Letter of Acceptance

A legally binding contract is established when the tender is formally accepted by or on behalf of the employer. It is usual practice for the engineer to prepare a draft letter of acceptance of the tender, setting out any agreed conditions or provisos which have transpired during the tender period, and to send this to the employer with his tender assessment or adjudication report. The letter should include a statement that all subsequent directions concerning the contract will be issued by the engineer, and that all further correspondence relating to the contract should be addressed to him, except where there is a contrary formal requirement in the contract. Copies of all correspondence should be sent to the engineer in order that he shall be kept fully informed of all matters relating to the contract.

Alternatively, the letter of acceptance may be issued by the engineer on the written authority of the employer, but making it clear that the engineer does not undertake any of the employer's obligations under the contract. In either case it is essential that the letter should be an unequivocal acceptance of the final offer.

Form of Agreement

It is also advisable for the Form of Agreement to be drawn up, signed and sealed by the employer and the contractor at this stage. A model form of agreement which can be used for this purpose is published in the ICE Conditions of Contract.[19] Prior to the completion of this document, the interests of the employer and the contractor are safeguarded by the Form of Tender, which normally states that the receipt by the contractor of a letter of acceptance from the employer will constitute a binding contract between the parties until a formal agreement is concluded.[5]

REFERENCES

1. Chartered Institute of Building, Estimating Information Service. No. 34. *Contractor Selection – a Guide to Good Practice* (Autumn 1979)

2. Economic Development Councils for Building and Civil Engineering. *The Public Client and the Construction Industries* (The Wood Report). HMSO (1975)

3. Institution of Civil Engineers, Association of Consulting Engineers and Federation of Civil Engineering Contractors. *Guidance on the Preparation, Submission and Consideration of Tenders for Civil Engineering Contracts recommended for use in the United Kingdom* (1983)
4. National Joint Consultative Committee for Building. *Code of Procedure for Single Stage Selective Tendering* (1977)
5. Institution of Civil Engineers. *Civil Engineering Procedure* (1979)
6. I.H. Seeley. *Quantity Surveying Practice.* Macmillan (1984)
7. C.K. Haswell and D.S. de Silva. *Civil Engineering Contracts: Practice and Procedure.* Butterworths (1982)
8. M. O'C. Horgan. *Competitive Tendering for Engineering Contracts.* Spon (1984)
9. R.J. Marks, R.J.E. Marks and R.E. Jackson. *Aspects of Civil Engineering Contract Procedure.* Pergamon (1985)
10. Banwell Committee. *The Placing and Management of Contracts for Building and Civil Engineering Works.* HMSO (1964)
11. Chartered Institute of Building. *Code of Estimating Practice* (1983)
12. R.S. Harrison. *Estimating and Tendering – Some Aspects of Theory and Practice.* Chartered Institute of Building, Estimating Information Service No. 41 (1981)
13. R. McCaffer and A.N. Baldwin. *Estimating and Tendering for Civil Engineering Works.* Granada (1984)
14. J.A.A. Sharp. *The Cost Estimate – A Need for Reconciliation –* Chartered Institute of Building, Estimating Information Service No. 40 (1981)
15. *Spon's Civil Engineering Price Book,* Edited by Davis, Belfield and Everest. Spon (1984)
16. J.J. Farrow. *Tendering: An Applied Science.* Chartered Institute of Building (1976)
17. D.A. Elliott. *Tender Patterns and Evaluation.* Chartered Institute of Building, Estimating Information Service No. 25 (1977)
18. C. Tassie. At the right price. *Building* (5 March 1982)
19. Institution of Civil Engineers, Association of Consulting Engineers and Federation of Civil Engineering Contractors. *Conditions of Contract and Forms of Tender, Agreement and Bond for use in connection with Works of Civil Engineering Construction.* Fifth Edition (June 1973, revised January 1979)

4 Site Organisation

This chapter examines site personnel, management and planning, site organisation, site layout, planning and monitoring of activities, resource scheduling, incentives and productivity, choice and maintenance of plant, and safety aspects.

SITE PERSONNEL, MANAGEMENT AND PLANNING

Contractor's Organisational Arrangements

A civil engineering contractor's organisation is usually subdivided into two main sections – technical and non-technical. The responsibilities of the technical section are likely to include the following:

(1) all civil engineering services, including the preparation of designs for temporary and permanent structures and the planning and programming of work;
(2) estimating and tendering, preparation of specifications and quantities, interpretation and application of contract conditions, and negotiations with the engineer;
(3) supervision of constructional work, monitoring of progress, preparation of reports, liaison with the engineer, purchasing of materials and equipment, and preparation of monthly valuations;
(4) quality control, research and development, site investigations and geotechnical processes;
(5) operation of central plant and transport depots with workshops and repair facilities, routine inspection of plant and equipment, and purchases and sales of plant; and
(6) staff training.

The non-technical section deals with a wide range of activities including:

(1) secretarial and legal matters;
(2) finance, accounts, audits, payments, cash and payroll checks;
(3) insurance, licences and taxation returns;
(4) orders, monitoring deliveries and checking invoices;
(5) cost records and analyses;
(6) plant and transport records and registers;
(7) general correspondence and records;

(8) labour relations; and

(9) staff training and development.[1]

Site Personnel

The contractor should send sufficient experienced technical and clerical staff to the site at the start of the contract to ensure that all necessary preparatory work is undertaken on the site. The full complement of site staff on a large project can be extensive and the principal members of staff and their main functions are now described.

Agent

An agent, or project manager as he is sometimes called, is normally appointed to control the contractor's site organisation. He is usually an experienced engineer and he is generally given wide discretionary powers by the contractor. In addition to sound engineering and contractual experience, he must possess good qualities of leadership and integrity. The agent's main duties are to ensure that the works are administered effectively and that construction is carried out economically, in accordance with the contract documents and that they satisfy the requirements of the engineer's representative.

Agent's Staff

The agent is supported by both technical and non-technical staff with their numbers and duties dependent on the size of the project. For instance, the personnel might comprise an engineer, general foreman, plant and transport foreman, cashier, timekeeper and storekeeper. Larger contracts require proportionately larger staffs. The allocation of duties will be influenced by many factors, including the locality and nature of the work, the amount of assistance from head office and the experience and capability of the available personnel. The allocation of duties should seek to ensure smooth and effective communication and the ability to introduce checks at critical points.

Figure 4.1 illustrates a contractor's typical site organisation based on an example in *Civil Engineering Procedure*, although it will be appreciated that there is no one universal system.

Sub-agents

Sub-agents are often appointed to control the various work sections of a large project, and they, in their turn, may have a number of section engineers responsible to them for the supervision of the actual operations. The direct control of labour and operation of plant and transport is usually undertaken by various grades of foremen.

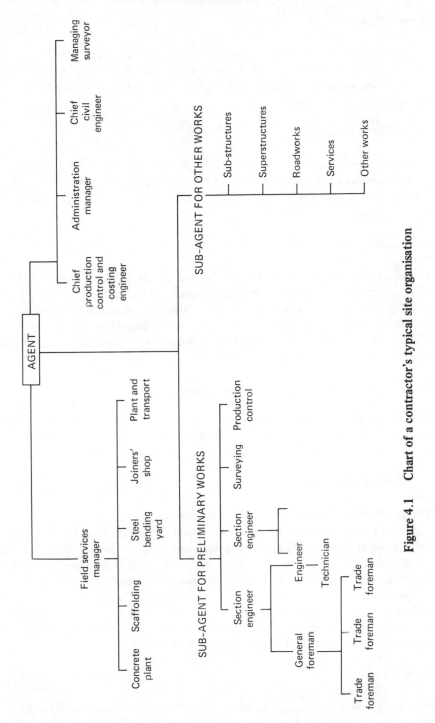

Figure 4.1   Chart of a contractor's typical site organisation

General Foreman

The general foreman or works manager directs the day-to-day distribution of labour to particular operations under sectional or trade foremen, to supervise the supply of materials and stores and disposition of plant, and to activate the necessary site communications.

Field Services Manager

The field services manager/engineer controls a number of departments which provide services needed for the effective execution of the project. These services often include a concrete batching plant, steel bending yard and plant and transport department.

Chief Civil Engineer

The chief civil engineer is responsible for the accuracy of the works through section engineers. He needs to check the co-ordination of drawings received from the engineer and then issue them to the relevant personnel in the appropriate sections of the project. He also carries out any local designs that may be required, particularly those relating to temporary works, and gives general technical guidance where necessary to personnel on the site.

Chief Production Control and Costing Engineer

The chief production control and costing engineer keeps routine progress records and costs and normally operates through departments controlled by sub-agents. He is often responsible for the routine measurement of work on site.

Administration Manager

The administration manager is responsible to the agent for the efficient administration of the non-technical personnel. He usually prepares detailed lists of duties for the guidance of clerical staff and devises checking and counter-checking procedures for cash transactions and stocktaking. He needs to be familiar with the standing instructions of the head office and to ensure compliance with them.

His department also controls the payment of wages through timekeepers and cashiers and the purchasing and checking of receipt of materials and components, in addition to the checking of accounts, insurance, safety precautions, site welfare and other matters relating to labour relations.

Section Engineers

Section engineers are usually engineers with experience of both design and field work who, although ultimately responsible to the engineer, report to the chief civil engineer on the accuracy and control of the civil engineering works. Each section engineer will liaise with his foreman to plan the work of his section and report on matters of detail to the managing surveyor or the quantity surveyor engaged on the project.

Managing Surveyor

On large contracts a managing surveyor may be employed to take measurements and check the quantities and value of completed work.

Site Management

Hillebrandt[2] has described how many agents or project managers have to make many *ad hoc* decisions because of the large number of inputs on to a construction site, the fact that each project is a one-off operation exposed to the vagaries of the weather, and the difficulties of managing large numbers of men possibly working together for the first time. Moreover, because of the diversity of conditions from one site to another, agents often have to take decisions based on personal knowledge and experience, without reference to senior personnel at head office. By comparison, most factory processes are based on clearly defined past practices and are often largely determined by the fixed plant and machinery.

Poor management is reflected in increases in total costs of various inputs to the construction process. For example, the cost of materials can rise through wastage resulting from bad storage, pilfering or lack of care in use. It can increase the cost of labour because of low productivity, poor workmanship necessitating rectification, and loss of time between activities arising from inadequate planning of the flow of operations. The cost of sub-contractors' work can rise because of poor planning, resulting in a delayed start on site and the subsequent submission of claims against the main contractor. The cost of plant can be increased because of low usage and inadequate maintenance.

The majority of agents or project managers on civil engineering projects are qualified professional engineers, normally members of the Institution of Civil Engineers. However, it was not until the nineteen-seventies that the Institution concerned itself to a significant extent with management, either in the formal training of civil engineers or as a subject for research or information dissemination to members. The major task of a civil engineer in charge of a site was generally considered to be the satisfactory execution of the works. However, other requirements of the employer, such as construction to a time and cost budget, have increased in importance and this has necessitated greater management control over resources.[2]

These developments have been assisted by the fact that civil engineers generally display more interest in management than architects, partly because their responsibilities have traditionally encompassed a wide range of professional skills on site, including the measurement and valuation of completed work, unlike building contracts, where a quantity surveyor almost invariably carries out these latter functions. However, the larger contractors are increasingly engaging specialists, such as quantity surveyors and accountants, on construction sites to perform functions previously undertaken by civil engineers. Civil engineers are recognising the need to acquire and demonstrate expertise in these areas, and they are now included in the curriculum of many civil engineering degree courses.

Personnel Management

Even although civil engineering construction is more capital intensive than building projects, the management of personnel remains a critical factor. Studies have shown that differences in productivity between companies, and even between departments within a company, are likely to result from differences in the way that personnel are managed.[3] Hence effective management of personnel is an important aspect of the project manager's duties.

Some construction companies have established personnel departments which are largely responsible for recruiting, employing and developing personnel. A possible structure for a personnel department in a large construction company is shown in figure 4.2 based on an example prepared by Fryer.[4]

One of the more important strategic tasks of personnel management is to continually analyse and reappraise the organisation's operations. This will assist in making decisions on suitable work structures and formal roles and relationships, to allocate responsibilities and define levels of authority. The personnel manager can contribute to the development of manpower forecasting and budgeting techniques. He can also play an important role in identifying the strengths

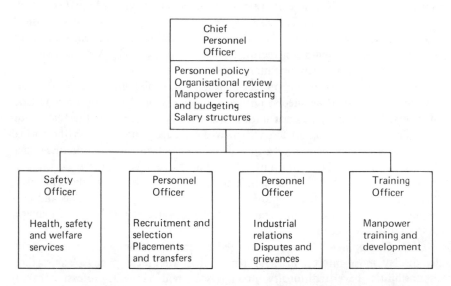

Figure 4.2   Typical personnel department structure

and weaknesses of the organisation and assessing the effects of social, legal, economic and other changes. Fryer[4] has identified the main tasks of personnel management as obtaining and retaining employees of suitable calibre, developing their potential and helping the organisation to manage people effectively.

Manpower Planning

The purpose of manpower planning is to maintain an adequate supply of suitably experienced personnel. It normally incorporates the following functions:

(1) analysing and describing posts and preparing job specifications;
(2) assessing present and future manpower needs;
(3) forecasting manpower supply and demand, and preparing budgets;
(4) developing and applying procedures for recruiting, selecting, promoting, transferring, and terminating the employment of staff;
(5) ensuring compliance with the requirements of employment legislation; and
(6) assessing the cost effectiveness of manpower planning.[4]

Forecasting future manpower needs is always a difficult assignment and should take into account:

(1) natural wastage due to retirement and employee turnover;
(2) promotions which create vacancies at lower levels;
(3) the company's plans for growth and diversification; and
(4) the availability of labour with the necessary skills in the required locations.[4]

A company needs to have readily available adequate information about its employees, including such aspects as age breakdown, skills analysis, succession plan, training needs and labour turnover analysis.

The project manager has to forecast manpower requirements for his site, taking into account the availability of the various categories of labour, including sub-contract labour, the need to avoid sharp fluctuations in manning levels and the overall resource pattern of the project, often illustrated diagrammatically in the form of a manning curve or manpower chart.[4]

The performance appraisal of employees should be subject to regular review to determine their likely potential, weaknesses, training needs and appropriate levels of pay. A primary aim is to make the company more efficient by increasing the competence of its employees and getting the best from them.

As the construction industry was forced to cope with a period of intensified depression in the late nineteen-seventies, personnel management and employees alike were affected by redundancies on a considerable scale. The Employment Protection (Consolidation) Act 1978 mitigated some of the financial effects of redundancy upon an individual, but the social and psychological effects can be demoralising. Therefore a company should pre-plan any redundancies to maintain good relationships with trade unions, employees and the public at large, and to retain the proper balance of skills for the completion of projects. The forward planning of manpower requirements should identify any possible excesses of manpower. If these occur, the company can consider various alternative strategies, including:

(1) ceasing recruitment other than for essential vacancies;
(2) transferring personnel from one area to another;
(3) retraining the existing labour force to match requirements;
(4) the phasing of any redundancies over a period of time; and
(5) the use of voluntary redundancy and the operation of a redundancy counselling service.[5]

SITE LAYOUT

Calvert[6] has described how a site layout plan should be prepared showing the proposed locations of all facilities, accommodation and plant to secure optimum economy, efficiency and safety during construction. A tidy site is the outward symbol of an efficient organisation. The following aspects deserve particular attention.

(1) Access

The requirements vary with the type of project and the stage of construction. Access from the public highway should desirably be duplicated, with short direct routes and one-way traffic to ensure a smooth flow of vehicles. Alternative transport by rail or possibly canal may be useful for bulk deliveries or long components. Temporary access ways may be constructed of hardcore, sleepers, concrete, proprietary track or transportable mats for mechanical plant. Wherever possible, advantage should be taken of any permanent works when siting temporary roads or hardstandings.

Even temporary road surfaces require adequate drainage and maintenance. Suitable cross-overs must be provided for tracked vehicles traversing metalled roads, and excavated material must be removed regularly from adjoining highways. On large sites, vehicle checkpoints may be established for security purposes.

Permission must be obtained from the local authority for access over or encroachment onto public footpaths, and the police must be notified if roads are to be closed or diverted. It may be necessary to erect fences or hoardings on site boundaries, to provide watching and lighting, to maintain rights of way and instal pedestrian walkways or vehicle tracks over trenches.

(2) Materials Storage and Handling

The principal objective is to minimise wastage and losses arising from careless handling, poor storage or theft, and to eliminate double handling or unnecessary transportation of materials and components. Suitable stores and compounds must be provided for tools and equipment, plant spares, and breakable or expensive materials and components. Racks should be erected for the storage of scaffolding, and stillages for oil drums, and storage areas set aside for bulk items such as bricks. Special attention must be paid to the storage of materials like cement which need to be protected against moisture, and goods which require careful stacking to prevent deformation, such as metal windows. Newly erected ancillary buildings can often be used for storage purposes and this can be facilitated by building them early in the construction programme. Sub-contractors' needs must be given adequate consideration and suitable space allocated for their huts and materials.

Calvert[6] has described how a site plan should show the locations and sizes of stacks of materials and components, the dates they are required, planned routes for distribution and final destinations. Economical methods of handling include the gravity feeds for aggregates, rubbish chutes, the use of packaged bricks and bulk cement, conveyors, off-loading gantries and stacker trucks.

Security measures must be given high priority. These can include locked buildings, substantial fences and gates, careful location of checkers' huts, the possible installation of a weigh-bridge, effective procedures for receipt and issue of stores, fire precautions, and the employment of a nightwatchman, guard dog or visiting patrolman.

Efficient distribution of the offices, stores and labour force, and an intelligently arranged programme, can ensure that materials are handled as little as possible and used in the order of arrival on site. For instance, if the roof steelwork is delivered to the site simultaneously with the main and secondary beams, the various steel members may become intermingled, resulting in a section of the work coming to a standstill while a search is made for a certain beam from a large pile of roof members. In like manner substantial disruption of the works could result from the mixing up of cleats and bolts.

(3) Administration Buildings and Other Facilities

The siting of administrative offices both for the contractor and the engineer's staff requires careful consideration, ideally to site them overlooking the works yet reasonably free from the noise and dust emanating from site operations. The size and layout of rooms is influenced by the number of staff to be accommodated, but the requirements of privacy for discussion, and accommodation for site meetings must be taken into account.

The requirements for the agent's office will normally be detailed in the specification. The furniture and fittings will usually comprise such items as a desk, at least three chairs, a telephone with extensions to other site offices, a long bench for reading drawings, a plan chest, two drawing boards with tee squares and set squares, filing cabinet, desk and drawing board lamps, post rack and chart board, safe, large lock-up cupboard, and wash-basin.

Site buildings should be capable of speedy erection and dismantling, and preferably be arranged in sections so that they can be extended or contracted according to the type and size of project. The sections making up the structure should be light yet strong. Adequate car parking facilities will also be needed for use by the site personnel.

Welfare facilities have become an increasingly important item on construction sites, necessitating adequate provision of mess-rooms or canteens, drying or changing rooms, and toilets.

(4) Plant, Workshops and Services

The choice of the most suitable type of plant is a matter of major significance and will be considered later in the chapter. There needs to be a good balance of plant and manpower for each site operation.

The number and size of workshops must be determined and their location chosen so that ready access to construction work is obtained without causing congestion of the site. The lines of both new and existing services must be considered when siting temporary buildings and roads. The installation of temporary diesel pumps, water and compressed air mains, electric power and telephone lines, and other services requires negotiation with the relevant authority and co-ordination with the general scheme of work. The level of the water pressure and/or electricity supply may affect the size, type and number of items of plant that can be used on the site.

(5) Special Problems

Some examples will serve to illustrate the type of problems that may be encountered on construction sites and their possible implications.
(i) *Confined sites* may entail parking restrictions, one-way traffic on approach roads, restricted timetables for delivery of materials, plant and equipment, and two-storey offices or gantries over public footpaths.
(ii) *Staged completions* involving the intermittent handover of sections of the works and the possible removal or relocation of temporary installations, involving detailed planning to avoid double handling and disruption.
(iii) *Adjoining property* can cause complications, particularly where demolition is involved, which can possibly result in the diversion of services, shoring or underpinning, and necessitate taking photographs as a precautionary measure in case claims for damage materialise. Permissible noise levels and methods of control are prescribed in *BS 5228: 1984: Noise Control on Construction and Open Sites.*

PLANNING AND MONITORING OF ACTIVITIES

Objectives of Programming

In order that civil engineering works shall be carried out efficiently, it is essential that they are carefully and properly planned in the first instance. Decisions will be required on a whole host of matters ranging from constructional methods, temporary works and plant, to labour, material and transport requirements all set against a background of time. The decision making process will also involve the consideration of alternative methods and the effect of each planned activity on the others.

To the contractor's agent, the programme shows his production target. It also plays an important role in the organisation of plant and labour, and the control of sub-contractors' operations and suppliers' deliveries, and it dictates the cash flow position of the contractor.

The contractor selects his method of working, both to satisfy the requirements of the contract, and to make the most economic use of labour, materials and plant. Excessive peaks and troughs in the use of expensive items of heavy equipment must be avoided at all costs. The contractor's agent must endeavour to secure continuous and concentrated use of all specialised plant and trades and to obtain optimum benefit of reusable items such as formwork panels, trench sheeting and scaffolding.

The programme consists of breaking down the contract works into a series of operations, which result in an orderly development by stages, from the temporary and preliminary operations, through to the completion of the permanent works. In the absence of a programme the work is likely to be carried out in a haphazard and disorderly manner. The programme should ideally be discussed and agreed by all concerned before work starts, to avoid confusion, delays and increased costs.[1]

Civil Engineering Procedure[1] categories programmes into three broad classifications.

(1) *Outline programmes* are prepared separately by the engineer and the tenderer. The engineer's programme is included with his report to the employer, while that of the tenderer is submitted with his tender.
(2) *A master programme* is prepared by the contractor as soon as possible after he receives instructions to start work on the project. It comprises a vital control document and shows the periods during which the individual work sections are to be carried out in order to achieve the correct sequence of operations and completion of the works within the contract period. All those in authority on the project, whether on the engineer's, contractor's or sub-contractors' staffs should be thoroughly familiar with the detailed contents of the master programme and should aim constantly to carry out the work as planned. The programme will also indicate the dates by which

detailed drawings will be required, the dates when the various sections of work will be completed, ready for use, or for the installation of plant by other contractors. It also provides valuable information relating to material, labour and plant requirements.

(3) *Detailed programmes* cover each section of the project and they dovetail into the master programme.

The contractor may also prepare programmes showing labour, plant and materials requirements respectively. Thus a labour programme, related directly to the master programme, will show the detailed requirements for all craft operatives and labourers throughout the contract period. The plant manager requires prior notice of the plant programme showing exactly what plant is required, where it is to go and for what period. Materials and components will be required on specified dates throughout the progress of the contract. The purchasing department must be informed well in advance of the quantities of materials required, and the dates when they will be needed on the site. This programme can be usefully employed to provide a warning routine, whereby suppliers are advised a week or two before the delivery date previously supplied and agreed, that the order still stands, and that delivery is required on the prescribed date.

The programme must be realistic and capable of fulfilment. For example, a programme which showed earthworks proceeding uninterrupted throughout the winter months with no allowance for stoppages resulting from adverse weather conditions is unlikely to be capable of achievement. Possibly the worst feature is where a programme is intentionally condensed to show an artificially early completion which may subsequently provide the basis for an exaggerated claim for delay. The engineer must examine the programme thoroughly with a view to detecting any errors in the logic or basic assumptions underlying the sequence and linkage of the various activities shown. Typical errors would be provision for the erection of beams prior to the supporting columns, insufficient time allowed for curing concrete, or delivery rates for materials which are incapable of realisation.[7]

Methods of Programming

The five principal methods of programming are each examined and compared.

1. *Bar Charts*

Bar charts are widely used and easy to understand, with bar lines representing the time period allocated to each operation, and the relationship between the start and finish of each activity is clearly seen. The timescale, usually related to a calendar but also often including the weeks in the contract period, is shown horizontally, and the activities listed vertically as illustrated in the relatively simple programme and progress chart in figure 4.4. Bar charts are, however, incapable of showing in detail a large number of interrelated activities and dependencies or whether one activity is delaying or about to disrupt another. Hence they have substantial limitations and do not permit a high degree of control. They are reasonably well fitted to cover a contract where a relatively small number of large scale, self-contained operations dominate the programme, as in the case of a road contract comprising mainly earthworks and carriageway construction elements with only limited structural content.

2. Elemental Trend Analysis

Elemental trend analysis or line of balance is not so readily understood as a bar chart but it highlights the importance of activity completion, production rates and relationships between selected activities. It also has a horizontal timescale and calendar, with cumulative output shown vertically. Bar lines representing the various operations are inclined at different slopes to indicate the rate of working. It is particularly suitable for repetitive work and strict operational sequencing and permits a high degree of control.

3. Network Diagrams

The need for greater efficiency and speedier completion of projects has resulted in the increased use of network analysis methods which can be subject to speedy revision by computer. This approach is very well suited to projects encompassing many activities which have to be linked together as interrelated steps in the completion of the project. For instance, the construction of an aircraft hangar involving piled foundations on site, extensive steel fabrication off site and special welding operations as the steel sections are erected in sequence.

The network is basically a project graph which clearly depicts the various operations which have to be performed to complete a project and their interrelationship. Each of these operations can often be carried out in a variety of different ways, by varying the amount of labour, number and size of gangs, working hours, plant employed and other inputs. The optimum combination normally represents the lowest cost, subject to availability of plant and personnel, and this can be determined by network analysis.

Peters[8] has outlined the following useful procedure for constructing a network diagram.

(1) Analyse the project into discreet activities.
(2) Determine the sequence and interdependence of activities and decide which activity must come first and what follows it.
(3) Assign time or cost values to each activity and, in some cases, both may be needed.
(4) Determine the critical path, being the critical sequence of activities through the network from start to finish.

Certain of the operations are critical, in that if they are delayed or prolonged, the completion date of the project will also be delayed. Other operations can be delayed without necessarily affecting the project completion date. Network analysis pinpoints the critical operations and enables the time leeway for all other operations to be determined.

Networks consist basically of circles, or sometimes squares or triangles and lines or arrows. The circle is an event in the life of the project and is not time consuming. It represents the point at which one operation is completed and another can start. The line or arrow shows the direction of work flow. One operation has often to be completed before another can commence, as illustrated in the following example.

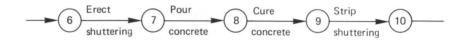

In this example the numbers represent events, for example shuttering fixed at 7 and concreting completed at 8, whereas the lines with arrow heads denote the actual activities. Broken lines also appear on networks to indicate dummy activities, which have no value in time or resource usage but merely show interdependency, as illustrated in figure 4.3.

In drawing up a network, three questions must be asked when considering each activity:

(1) What activity must immediately precede this operation?
(2) What activity can immediately follow this operation?
(3) What activities can be taking place concurrently with this operation?

In the critical path method the time factor is very important, and the estimated time for each activity is shown on the network. Earliest event times can then be calculated and are often entered on the diagram in squares or circles, adjoining the event numbers. Subsequently, latest event times, which still permit the completion of subsequent events within the overall project time, are added. The critical path shows the chain of critical activities along which any delay will prolong the completion time, and this path is usually indicated by a thick line.

The spare time or leeway available in activities off the critical path is termed 'float'. The total float is the difference between the latest finishing time and the earliest starting time, less the duration time for the activity. Once the critical path and overall project time have been established, it is necessary to decide whether the project requirements are satisfied. Where an improved schedule is requested, the critical path will be the first element to be examined.

Critical path analysis of civil engineering projects can be conveniently broken down into six main processes:

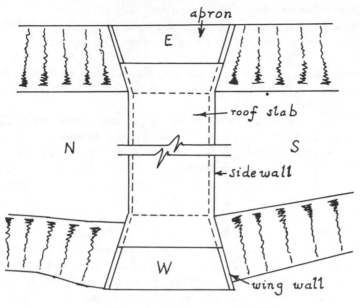

PLAN OF CULVERT

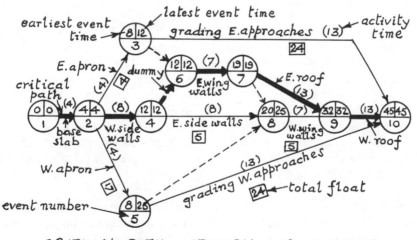

CRITICAL PATH NETWORK FOR CULVERT

Figure 4.3 Critical path network

(1) project study;
(2) planning and development of an arrow network for the project;
(3) programming — putting the plan on a time basis;

(4) determining project duration – critical path and floats;
(5) reappraisal and possible revision of network and programme;
(6) preparation of programmes for management control purposes.

A typical critical path network analysis is shown in figure 4.3, encompassing the construction of a reinforced concrete culvert, with the activities listed for information purposes.

The numbering of the events follows the sequence of operations on the site, commencing with the construction of the base slab and finishing with the roof at the western end of the culvert. Earliest and latest event times and event numbers are shown in circles. For instance, the earliest event time for event nr 5 (completion of west apron) is $4 + 4 = 8$ days. The latest event time is the same as for event nr 8 (25 days), ready for the start of the west wing walls. The dummy activities are shown by broken lines and the critical path in a thick line. The critical path consists of the chain of events wherein the earliest and latest event times are the same, and there is, therefore, no spare time or float. The total construction period of 45 days will need considering in relation to the programme for the complete project, and if this period proves excessive it will be necessary to review the constructional methods with a view to reducing the time required for the activities on the critical path.

Floats may be of four different types:

(1) *Total float* is the difference between the latest finishing time and the earliest starting time, less the duration time for that activity.
(2) *Shared float* is common to more than one activity and if used by one activity is no longer available to another – for example, activities 4 to 8 and 8 to 9, where a shared float of 5 days is available.
(3) *Independent float* operates when an irreducible, but longer than necessary time is available for an activity – for example, activity 5 to 10, where the independent float = $45 - 25 - 13 = 7$ days.
(4) *Free float* is the difference between the earliest finishing time and its succeeding earliest event time for any activity – for example in activity 8 to 9, where the free float = $32 - 20 - 7 = 5$ days, and it cannot be used by succeeding activities.

It is customary to prepare a work schedule, indicating all the operations shown on the network analysis with their earliest and latest starting and finishing times, total, independent and free floats and critical path activities.

4. Precedence Network Diagrams

Precedence network diagrams resemble a collection of linked boxes with visual emphasis on activity descriptions. Activities can be grouped by work section and location and construction sequence is highlighted. The technique is ideal for large complex projects, but reference to calendar time is only indirect and skill is required in both presentation and interpretation.

5. Linked Bar Charts

Linked bar charts retain the visual benefits of bar charts with increased emphasis on dependencies. This takes the form of vertical links between the completion of one activity and the start of another. The greater emphasis on co-ordination and construction sequence allows the technique to be used for more complex projects than the normal bar chart. However, the float concept is generally missing and there is a limit to the amount of linking that is possible.

Progressing

Progressing consists of taking systematic steps to ensure that the programme is followed as closely as possible. Unless this monitoring procedure is adopted, the value of the programme will be largely lost. Lack of progress will necessitate some modification of the programme. It is essential to introduce an adequate recording procedure to ensure that a thorough, accurate and regular check is made of the work executed under the work section heads or operations contained in the programme.

A relatively simple programme and progress chart covering a small roadworks contract is illustrated in figure 4.4 to show a common approach to this activity. The contract period is ten weeks and the chart indicates both the programme and progress at the end of the sixth week. Cumulative weekly quantities of work planned and executed are inserted for each of the three main sections of work. Each thick black line indicates the quantity of work completed and the hatched sections show the work still remaining to be done. The thin black line shows the period throughout which the particular operation has been performed. On occasions the average weekly targets of completed work in each section are also indicated as shown in figure 4.4.

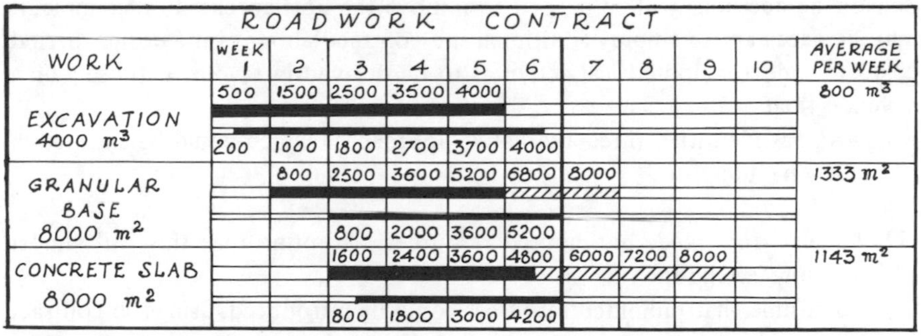

WORK	WEEK 1	2	3	4	5	6	7	8	9	10	AVERAGE PER WEEK
EXCAVATION 4000 m³	500	1500	2500	3500	4000						800 m³
	200	1000	1800	2700	3700	4000					
GRANULAR BASE 8000 m²		800	2500	3600	5200	6800	8000				1333 m²
			800	2000	3600	5200					
CONCRETE SLAB 8000 m²			1600	2400	3600	4800	6000	7200	8000		1143 m²
			800	1800	3000	4200					

(ROADWORK CONTRACT)

Figure 4.4 **Programme and progress chart**

The position at the end of the sixth week is that the excavation work was completed half a week behind schedule, the laying of granular base is a week behind programme but this presents no difficulties, and the concreting work is half a week in arrears, but provided the present rate of output is maintained, the project can be completed within the contract period. Programme charts usually contain far more items than the one illustrated; also there is usually a preliminary and temporary works item at the start, and a clearing up item at the finish, each possibly occupying a week or more. Additionally the programme may incorporate information on plant and labour requirements for each operation.

RESOURCE SCHEDULING

Resource Allocation

Lock[9] has described how a network cannot normally, on its own, identify the volume of resources needed at any given point in project time. When the network is prepared, no account is taken of the resources which will be available. The start of an activity is usually assumed to be dependent upon the completion of previous related events, and not on the availability of suitable operatives and plant. In reality, no project is carried out with unlimited resources and in completing the project plan it is necessary to have regard to resource limits.

The use of float time will often be of considerable help. For example, if an overlap of activities resulted in the need for two tower cranes on a site, this would necessitate some re-programming to reduce the tower crane requirements. On large projects the procedure for achieving this is very complicated and it is advisable to use a computer to perform these calculations, as described later in the chapter.[10]

Resource scheduling is an essential follow-on step from the initial planning network. The results of time analysis are used to determine priorities when different activities compete simultaneously for the same limited resources. Scheduling decisions will be needed to remove the bottlenecks. For example, it may be necessary to employ additional sub-contract labour over a critical period or to provide for non-critical activities to be delayed in favour of those which have less float.

Lock[9] has identified three alternative approaches which could be adopted in resource scheduling.

(1) To plan for fixed, but limited resources, accepting that this will almost certainly extend the timescale.
(2) To assume that unlimited resources could be introduced, using sub-contract labour as necessary to prevent slippage in the overall programme. The scheduling must be carried out realistically to remove all peaks and troughs

in scheduled requirements which are not imperative to the effective completion of the project on time.

(3) To adopt a compromise solution, allowing a restricted programme extension while, at the same time, also using limited additional resources.

Resource Scheduling by Computer

Various attempts have been made to apply resource allocation to networks, and many commercial computer programs of network analysis have been expanded to include this facility. However, the use of such enlarged programs has often proved restrictive. The program writers have aimed at using the computer to list and summarise the various resources and also to carry out 'resource smoothing', to ensure the most economical use of resources while still keeping within the prescribed overall program period. The procedure often takes the following pattern as described by Lester.[11]

(1) A network is drawn showing all activities.
(2) Durations are added.
(3) The activities are then listed.
(4) The resource for each activity is added to the list.
(5) The quantity of each resource, either in total or per time unit, is added to the list.
(6) The whole list is punched on card or on tape and fed into the computer.
(7) The computer calculates the floats and critical paths.
(8) The resources are aggregated for every time period.
(9) When the resources in any time period exceed a previous input resource availability, the computer extends the duration of activities with float so that some of the required resources are carried over into another time period(s), thereby reducing the total in the time period for which insufficient resources were available.
(10) This repetitive operation is continued by the computer until the resources are 'smoothed' and no significant peaks or troughs remain over the period of the program.

The computer can be programmed to perform the smoothing operation of each resource until no excess requirements remain in any time period, but this may extend the overall duration of the program. Alternatively, the facility normally exists to stop the resource smoothing operation when the previously calculated or pre-determined program/completion date has been reached. Then the excess resources are listed to enable management to take further action.[11]

The greatest advantage to be gained from the use of a computer lies in its ability to process large volumes of data with low risk of error and in a relatively short length of time. This enables a project manager to produce his schedule with a speed and accuracy which would not otherwise be possible. If a change in project circumstances is foreseen, it is relatively easy to change the data supplied to the computer and a revised schedule can be prepared within a few hours.

Lock[9] has described how the ability of the computer to sort and collate data is invaluable. It is possible, for example, to request a report which lists all activities, with dummies excluded, arranged in the sequence of their scheduled start dates. Resource requirements can be printed out on a day-by-day basis. Cost control data can be linked to the schedule and it is also possible to schedule several projects simultaneously, within a common pool of available resources.

A project manager must be sure that he has:

(1) a computer installation of sufficient capacity;
(2) a suitable program or software package; and
(3) systems support engineers who can answer questions or resolve problems arising from the use of the computer or the program.[9]

Microcomputers can be operated in ordinary office conditions by non-specialist personnel, such as the members of a project planning and scheduling team. However, microcomputers may not be able to cope with the larger networks, necessitating the use of the more expensive minicomputers. It is advisable to gain considerable experience as a user, possibly involving some form of sharing or buying time on a suitable computer before deciding on which particular combination of hardware and software to purchase.

Lock[9] has described how it is possible to rent, lease or purchase an installation consisting of the input and output parts of a computer — an input keyboard normally being part of a visual display unit (VDU) and a line output printer. These can be connected to a remote computer at a bureau by means of an ordinary telephone line, provided a suitable device, called a modem, is installed with the terminal. With this system the user does not require an expensive central processor or the tape and disc units needed to store the data and program instructions. All he has to do is dial the computer from an ordinary telephone and then throw a switch to connect up the terminal. Queueing may arise where other users simultaneously require the bureau facilities but delays are unlikely to exceed a few seconds. Furthermore, the user is not restricted to a single bureau, since any service offering remote facilities can be obtained by dialling the appropriate telephone number.

A simpler and cheaper type of remote installation is provided by a Teletype terminal, which operates over normal telephone lines using a modem. However, Teletype operation is slow and the number of characters obtainable across the width of a page is reduced. These machines have only limited application for network analyses.

Developing a suitable network analysis program is expensive since it involves many man-hours in preparing the initial specification, making a systems analysis study, writing the program and testing it. A project manager wishing to use a computer for the first time will probably make use of an existing program from amongst the large number of commercially available programs. The first step would be to write a summary specification of project-scheduling needs. It will be possible to examine the brochures provided by the program suppliers and to eliminate those which are unsuitable. The final choice will depend on the suitability of the program and its convenience in use.

COSTING AND ACCOUNTING ARRANGEMENTS

Site Staff

The *timekeeper* records the times of individual attendances and absences of operatives and compiles pay sheets. In conjunction with the foreman and gangers, he allocates times to the various project operations for costing purposes.

With regard to purchasing, some contractors negotiate and place all orders through head office buyers, while others delegate to regional offices controlling a number of contracts or directly to a specific contract where it is of sufficient size. A *buyer* requires wide experience of materials and stores in general use for constructional work, and of the customary quality classifications and price customs. He also needs to have a sound knowledge of commercial law, and particularly of the provisions of the Sale of Goods Act. He works closely with the engineering staff who normally determine the materials requirements to meet the programme. He collaborates with the storekeeper in pursuing suppliers who lag behind with deliveries or where earlier despatches are to be negotiated.[1]

The *storekeeper* checks and takes into custody all materials received, and subsequently issues them under a recognised system of authorisation. On some projects the storekeeper is also responsible for maintaining the plant register.[1]

The duties of an *invoice or accounts clerk* in a site organisation will vary with the accountancy system operating at the contractor's head office. His main function is to ensure that checked particulars of expenditure and receipts are readily available for entry under prescribed headings in the contractor's main account books at head office.[1]

Accounting Arrangements

An ICE publication[1] lists the following functions of an accounts system:

(1) assembling expenditure and liabilities incurred in the form and detail required by management;
(2) summarising relevant information to aid management in assessing present, and forecasting future, financial positions;
(3) providing up-to-date accurate financial statements at regular intervals and realistic estimates at interim periods if required; and
(4) preparing the normal accounts and returns of the business in conformity with statutory requirements.

The accounting method adopted will need to be subdivided into those duties to be performed on the site and those to be carried out at head office. The site duties normally include:

(1) accounting for all cash payments relating to wages and site expenses;
(2) estimating weekly cash requirements in advance; and
(3) checking and classifying invoices against deliveries of materials and stores.[1]

Cash Flow

When an employer contemplates investing funds in a large project spread over several years, he will wish to know the sum of money involved and the timing of payments. The employer needs to make arrangements with his banker or other financial source for the required sums to be available as required. Where projects are expected to make an eventual profit for the employer, the extent and timing of revenue receipts need predicting as part of the project appraisal, in order to assess the commercial viability of the project.

Various techniques are used in project appraisal but probably the most commonly employed is discounted cash flow (DCF). This technique permits all cash outflows (expenditure and taxation) and inflows (profits or savings) to be considered, with each individual component discounted to its future value. For example, £250 000 spent today is worth more than the same sum spent in one year's time. If the current interest rate is 10 per cent, then £225 000 can be invested now to accumulate to £250 000 in one year's time. In this example, 10 per cent is the discounting rate and £225 000 is the nett present value (NPV) of the final cost.[9]

In order to prepare a cash flow schedule, all relevant construction items with their associated costs have to be listed and subsequently entered on a table showing the timescale, often on a quarterly or monthly basis.

The contractor will need to monitor project costs very closely. At site level, the cost control often encompasses standard costs and variance analyses normally operated on a weekly basis. Head office management usually prepare cost/value reconciliations which are statements of the total costs and values relating to each project normally updated at each monthly valuation. A more detailed examination of cost control by both the contractor and the engineer is contained in chapter 7.

PRODUCTIVITY

The quantity of labour required to perform a specific item of work is described as the labour output. Contractors record details of outputs achieved on past projects and under the prevailing conditions. These provide valuable data for estimating purposes and determining suitable performance standards. The time taken to carry out the same item of work can vary from firm to firm, gang to gang, and area to area. It will be influenced considerably by the organisation of the firm and the familiarity of the gang in performing the particular task.

For example, the time taken to build a square metre of brickwork, one brick thick, should be calculated from the amount of brickwork built over a prescribed period of time or from the time taken to build a predetermined amount of brickwork. These calculations take into account all working time which is lost for a variety of reasons.

Milne[12] has described how method studies may be undertaken to determine performance standards and whether or not the most effective method is being used. For instance, a competent craft operative performing an item of work in an efficient manner under proper supervision would be allocated a rating of 100, and this is referred to as the standard rating. Work done at the standard rating with appropriate allowances for relaxation, personal needs and the like, gives a level of performance which is referred to as the standard performance.

In practice, craft operatives are unlikely to achieve the standard performance unless they are provided with a suitable incentive. A normal rating for a competent craft operative not working under bonus arrangements is likely to be in the range of 75 to 80. A rating of 60 would indicate a slow rate of performance with the operative showing little interest in his work. On the other hand a craft operative working on an incentive bonus scheme and producing good quality work to a rating of 125 would be displaying a high level of proficiency.

To maintain effective control of productivity, it must be forecast under any given set of conditions, and variances from the forecast monitored. In the future, predictive models are likely to be developed assisted by the increasing use of microprocessors.

As indicated previously, one of the main aids to increased productivity is the operation of incentive schemes, which by 1979 were being applied to about one-half of the operatives in the construction industry, with a high rate of take-up by the larger firms. Incentive schemes aim to secure greater productivity for the employer with employees receiving additional payments for their increased output. The basis for determining a target must be some readily ascertainable standard, normally obtained from the analysis of data on site. This deserves careful consideration because one factor that is difficult to determine is the rate at which the operatives were working when work was carried out.[13]

Oxley[14] has identified the main features of a good incentive scheme as:

(1) The amount of bonus paid to the operative is in direct proportion to the time saved and there is no limit to the amount that can be earned.

(2) Targets are issued wherever possible for all operations before work commences. The extent and nature of the operation should be explained clearly. Incentive packages are grouped to enable gangs to complete them in a relatively short time.

(3) Targets are not altered during an operation without the agreement of both parties.

(4) Systems used for calculating bonus earnings are clear to all concerned.

(5) Arrangements are made for dealing with time lost due to reasons outside the operative's control.

(6) Bonus payments are made weekly.

Productivity cannot, however, be considered in isolation from quality control and safety aspects. Quality control is concerned with defining the levels of acceptable imperfection, or tolerances, and in ensuring that prescribed minimum standards are achieved, and is examined in detail in chapter 5. Safety aspects are considered later in this chapter.

Skoyles[15] has described how not all materials delivered to construction sites are used for the purposes for which they are intended, and how contractors frequently use more materials than those for which they receive payment. These materials are either lost or are used in the construction process in ways that are not recognised by estimators. These differences have never been clearly defined and are collectively referred to as waste. For example, the estimator's normal wastage allowance for bricks and blocks is 4 to 5 per cent, whereas overall waste on the sites investigated by the Building Research Establishment ranged between 8 and 12 per cent.

Waste can be classified under two broad headings — direct waste (a total loss of materials which can occur every time materials are handled, moved, stacked or stored) and indirect waste (a monetary loss related to materials and sometimes to the way they are measured, and which can arise from substitution, production and operational loss and negligence). These losses highlight the need for engineers to pay more attention to ensure that design dimensions match the sizes of materials available on the market, for estimators to take a fresh look at estimating practice, and for site management to exercise more effective materials control.

PLANT USAGE, MAINTENANCE AND COSTING

Plant Requirements

The design of construction plant is highly specialised and demands a wide experience of the performance of machines in diverse conditons and the ability to make use of new materials and techniques. Construction machinery has to withstand severe weather conditions and exceptional stresses. Hence it needs to be tough, durable and resilient, easy to dismantle and reassemble on overhaul and capable of movement from one site to another.[1]

When selecting plant, the contractor should give special consideration to the following criteria:

(1) It must be the right type of machine for the work in hand.
(2) It must be able to carry out the task efficiently, economically and, if necessary, continuously.
(3) There must be sufficient work on the site to keep it occupied.
(4) There must be sufficient operatives and ancillary plant to maintain optimum output.

For example, an excavator must have sufficient earthwork to ensure optimum use and it must be adequately serviced by vehicles running between the site and the tip.

Plant must be maintained in full repair and should be kept on site for as short a period as possible. This applies particularly to plant hired by the contractor since he will have to pay for it whether it is in use or not.

Holmes[16] has listed the following important factors to be considered when selecting plant for a project:

(1) workload to be undertaken;
(2) time allowed in the construction programme for the work;
(3) capabilities of the machine and the various tasks which it can perform;
(4) transportation costs involved; and
(5) maintenance requirements and facilities.

Other factors may include the nature of the project and site, obstructions, boundaries, access and noise limitations.

With earthmoving plant, a prerequisite is often the preparation of a mass-haul diagram which shows the distances and direction of haul and gradients calculated to equalise the cut and fill. The selection of plant is largely dependent on the following factors:

excavating plant — quantities; type of soil; likely weather conditions; speed of removal; depth of dig; method of disposal.
transporting plant — quantities; distance to be moved; site conditions; tip arrangements; speed of excavation; size of excavating bucket; turnround time.
placing plant — method of transporting and compacting; quantities; likely weather conditions; required finish.
compacting plant — nature of material; contract requirements; depth of fill; likely weather conditions.[16]

Nature and Use of Plant

Much of the plant used on a civil engineering construction site is concerned with site handling, and includes cranes, mixing plant, vehicles, concrete pumps and hoists. Hence, efficient construction is largely dependent on selecting handling equipment that is best suited, both operationally and economically, for the project.

Illingworth[17] has classified contractors' site handling plant into three broad categories.

(1) Linear, embracing rope and pulley and all types of hoist.
(2) Two-dimensional, consisting of all wheeled transport, conveyors and aerial cableways.
(3) Three-dimensional encompassing all cranes, concrete pumps, fork lift trucks, telescopic handlers and aerial work platforms.

Each of these broad categories of plant will now be considered in outline, but more detailed information on civil engineering plant is contained in the ICES Guide.[18] The latter publication covers cranes, dozers, dumpers, excavators, piling equipment, pumps, compactors, rollers, scrapers, tractors, trenchers and wheeled loaders.

Linear Methods

Linear motion takes place in a vertical plane and the relative inflexibility of hoists almost inevitably means that they require backing up with secondary handling equipment to provide a complete handling operation.

The one exception arises when concrete, for example, has to be placed in tall structures of limited cross-section, as in tall bridge columns. In these locations an automatic discharge concrete hoist can be used with a short chute to discharge directly into the forms, as used on the Erskine Bridge in Glasgow. Although even then the hoist had to be fed from a ready-mixed concrete truck, which is basically a secondary distribution method.

Hoists of all kinds are relatively cheap to hire and operate, although the requirements of the *Construction (Lifting Operations) Regulations 1961* give rise to additional installation costs. The main value of the hoist is secured in three different situations.

(1) It provides the cheapest overall solution when costed in conjunction with secondary handling systems.
(2) Where the fast vertical speeds available are used to make secondary handling more productive, as when distributing concrete in tall structures by a tower crane.
(3) In tall chimneys, towers or deep shafts, where both men and materials have to be handled, often in confined spaces.[17]

Two-dimensional Methods

In the construction industry, the main use of wheeled transport is to distribute concrete, spoil and fill. A more limited category of plant consists of general purpose vehicles, such as flat bed trucks, tractors and trailers, and dumpers. Dumpers (up to 1 m^3) are used with small site concrete mixers, true-cretes (2 to 3 m^3) with site central batching plants, and truck mixers (4 to 8 or 10 m^3) for ready-mixed concrete.

Whichever type of wheeled vehicle is being considered, it will always form part of a chain of events to provide a complete handling cycle. Hence, its capacity and cost will need to be matched to the chain and the overall workload requirements. Furthermore, wheeled handling systems require adequate access roads otherwise slower haulage speeds and excessive lost time and maintenance costs could result.[17]

Other two-dimensional items of plant include conveyors, which are often used for moving excavated material over long distances, and aerial cableways which are particularly suitable for handling formwork and placing concrete on dam construction, but entail high capital costs.[17]

Three-dimensional Methods

Three-dimensional methods offer the widest scope and encompass cranes, concrete pumps, fork lifts, telescopic handlers and varieties of aerial work and materials handling platforms. Of all the plant used for site handling, cranes are probably the most important. They can be separated into a number of clearly definable groups of plant, namely: derricks; excavator conversions; lorry or wheel mounted; hydraulics; tower; and portal gantry.

In planning crane operations adequate care must be taken to safeguard the public. In particular, compliance with the *Construction (Lifting Operations) Regulations 1961* and *CP3010: 1972 — Safe Use of Cranes* must be ensured.

Earthmoving Plant

Horner[19] has classified earthmoving equipment into four broad categories:

(1) Excavation: rippers, drill and blast; impact hammers; hydraulic breakers; graders; and skimmers.
(2) Excavation and load: dragline; face shovel; forward loaders; grab; back acters; and bucket wheel excavators.
(3) Haul and deposit: dumpers; dump trucks; lorries and conveyors.
(4) Excavation, load, haul and deposit: dozers; tractor drawn scrapers; motor scrapers; and dredgers.

The relative ease of excavation is related to the power and capacity of the plant and the location and accessibility of the material to be excavated. For example, a motor scraper excavates material from below itself and requires the material to be of sufficient extent, of limited strength range and lying at a low gradient, while back acters perform best in removing material below the level of their tracks and can operate on ground of lower strength than motor scrapers.

Forward loaders and face shovels generally excavate material above the level of their tracks and a dragline can excavate material at distances of about 20 m from the machine, below the level of its tracks.[19]

In order to decide the types of earthmoving plant to be used, the work should be carefully analysed. The output rates of the various suitable and available combinations of plant or plant teams and their operative costs can then be determined. The output rates can be calculated from the manufacturers' data or, more reliably, from the contractor's own records based on experience on sites.[19]

Plant Maintenance

The operation and maintenance of plant on the site is supervised by plant and transport managers supported by foremen and mechanics, preferably with experience of site conditions. The continuous operation of plant and transport depends on good site maintenance, and this requires well equipped workshops, adequate supply of spare parts and efficient storekeeping.[1]

The maintenance of plant can be conveniently subdivided into two categories:

(1) regular overhauls; and
(2) day to day servicing while the plant is in operation.

The object of the overhaul is to strip down the whole or part of the machine and to replace parts which are badly worn and unlikely to operate satisfactorily until the next overhaul, which could result in a breakdown of the machine. This activity is sometimes referred to as planned preventive maintenance.

Servicing consists of inspecting the machine at daily or weekly intervals and taking the appropriate action to keep the machine in good working order. It includes such activities as cleaning and lubricating, checking water, oil, fuel, lights, tyres and brakes, tightening loose nuts and bolts, carrying out necessary adjustments and minor renewals, and taking precautions against frost.

The adoption of a standardised routine for periodic inspections is vital. Good maintenance aims at avoiding delays due to breakdowns. Any delays that do occur should be carefully recorded and the causes investigated to ascertain whether the breakdowns result from defective design, materials or manufacture, or poor maintenance or misuse.

Plant records should be maintained for every item of plant and should show the amount of fuel consumed, hours of use, servicing, overhauls and replacements. From these records it is possible to predict the length of working life of the various parts of the machine prior to renewal. They will also indicate when a machine should be written off as being no longer economical to operate, and will assist in determining or amending the necessary overhaul frequency.

Plant Costs

A contractor should make the decision to purchase plant, as opposed to hiring it, based upon an economic analysis of the alternatives. Included in this analysis will be such factors as availability of cash for a down payment or outright purchase, current interest rates on loans or instalment purchase plans, current hire rates, tax benefits, and expected use and anticipated life of plant. Ownership of plant can be a drain on resources if the plant is underused, subject to frequent breakdown or becomes obsolete.[20]

As a general rule, large and expensive items of plant are best hired as required, usually for relatively short periods and achieving a high rate of output. General items of plant which have a high usage factor are usually better owned. However, on occasions, the duration of a contract and the amount of work involving the use of plant may create a situation where it could be advantageous to purchase all new plant and dispose of it on completion of the contract.[16]

Calvert[6] has shown that when comparing alternative plant costs, it is necessary to consider the implications of each of the following factors:

(1) The working life of an item of plant will depend on its quality, strength and durability and how it is used and maintained. Past experience may provide a good guide.
(2) The extent of use of an item of plant requires accurate prediction in order to assess an hourly plant rate for charging to contracts. Furthermore, hire charges will be levied upon the time it is available for use on the site and not the actual time worked.
(3) Market depreciation must be calculated so that the value of the item of plant is progressively reduced and a replacement fund accumulated from which to purchase a new machine when the existing machine becomes obsolete and possesses only a minimum residual or scrap value. Replacement costs should incorporate an allowance for inflation. Obsolescence may also occur sooner than anticipated if new models are produced with lower operating costs and increased outputs.
(4) It is necessary to include interest charges on the capital expended to finance the purchase of plant.
(5) Allowance must be made for maintenance and repairs to cover the cost of labour and materials used in servicing and the replacement of worn parts.
(6) The all-in labour costs of operating the machine will be converted to an hourly working rate of the plant.
(7) Allowance will also be needed to cover the cost of fuel and oil to operate the machine.
(8) There may also be variable costs such as the transporting of the machine to

and from the site and assembly and dismantling, together with any ancillary work.

(9) An annual addition is needed for insurance, licences and any administrative overheads, and possibly profit.

The main advantages of hiring plant are that it increases the contractor's liquidity; enables him to hire the most efficient and latest plant for the particular project; and permits greater control of costs. The principal disadvantage stems from the possible non-availability of plant when required.

SAFETY ASPECTS

General Background

Construction sites often create potentially dangerous situations and about 100 persons are killed on them in the United Kingdom each year. Over 40 000 construction accidents were reported to HM Factory Inspectorate in 1982 and the safety statistics give considerable cause for concern.

Some activities are more vulnerable than others. For example, steel erection and demolition account for many fatalities and serious accidents. Many site injuries result from operatives falling from structures or being hit by falling objects. Many others are caused by the misuse of mechanical plant and site transport.[4]

Statutory Requirements

The construction industry falls within the ambit of the *Factories Act 1961* and the *Construction Regulations 1961 and 1966*. Subsequent legislation in the form of the *Health and Safety at Work, etc. Act 1974* provided the legislative framework within which to promote, stimulate and encourage high standards of health and safety at work. It aims to persuade all those involved in the construction industry, including employees, to promote an awareness of safety during the construction process, and to do all that is necessary to avoid accidents and occupational ill-health.

The main objectives of the Act are:

(1) to secure the health and safety and welfare of persons at work;
(2) to protect persons other than those at work against risks to health and safety arising out of work activities;
(3) to control the keeping and use of explosives and highly flammable or otherwise dangerous substances and generally prevent the unlawful acquisition, possession and use of such substances; and
(4) to control the emission into the atmosphere of noxious or offensive substances.

Employers have detailed responsibilities:

(1) to develop systems of work which are practicable, safe and have no risk to health;
(2) to provide plant to facilitate this duty, and this general requirement is to cover all plant used at the workplace;
(3) to provide training in the matter of health and safety; employers must provide the instruction, training and supervision necessary to ensure a safe working environment;
(4) to provide a working environment which is conducive to health and safety; and
(5) to prepare a written statement of safety policy and to establish an organisational framework for carrying out the policy; the policy must be brought directly to the attention of all employees.

However, employees also have the following specific duties:

(1) to take care of their health and safety and that of other persons who would be affected by acts or an omission at the workplace; and
(2) to co-operate with the employer to enable everyone to comply with the statutory provisions.

Practical Applications

Safety Officers

Contractors undertaking extensive civil engineering projects employ a full-time safety officer(s). The role of safety officers varies from firm to firm, depending on the size of the organisation and the type of work undertaken. The duties could include formulating safety policy; advising management on safety matters; drafting safe working procedures; investigating accidents and preparing and analysing safety records; safety training and propaganda; inspecting sites; and monitoring work of safety committees. Safety officers need to work closely with management and the trade unions to ensure that no aspects of safety are ignored.[5]

Earthworks

The following precautions should be taken to eliminate or reduce the accidents resulting from earthworks:

(1) The failure of temporary slopes is a common cause of accidents and all batters should be cut to a safe angle or shoring constructed.
(2) Neither heavy plant nor excavated material should be sited near the edge of batters.

(3) When excavating from the base of a working face, care should be taken to ensure that the face is not overhanging or excessively high.
(4) Open overnight excavations that could be a hazard to the public and site personnel, should be protected with barriers and hazard warning lights.
(5) Explosives should be handled only by experienced personnel, satisfactorily stored and adequate precautions taken when blasting.[19]

Site Handling

Care must be taken to ensure that plant operators and banksmen, are properly trained, competent supervision provided, and that all involved in a handling operation are adequately briefed as to requirements.[17]

Tunnelling

Safety in tunnelling is the subject of BS 6164.[21] There are special hazards arising from the nature of the work, the confined space and problems of access. Contingency planning should be undertaken prior to construction to prevent unexpected hazards becoming catastrophes. Natural hazards arising from ground conditions include collapse, inundation and gas.

The most vital safety measure is to be aware of these dangers at all times and to take appropriate corrective action before serious problems occur. A first line of defence is to restrict the excavated area and to ensure the availability of materials for immediate support.[22]

Sewers

Men working in foul and surface water sewers, manholes, siphons and other underground confined spaces need to be highly trained and to follow closely standardised procedures because of the risks involved from flammable, asphyxiating and poisonous gases, and possible flooding. All personnel must be fully equipped with all appropriate protective and safety clothing and equipment.

The standard precautions are extensive and range from erecting barriers and warning signs around three manholes, opening them up and checking for unusual conditions. The nearest telephone should be checked to ensure that it is operational and a strict procedure followed on entry to the sewer by a team with lookouts posted at both top and bottom of the access manhole. In the event of possible rain the men are recalled and in situations when the normal five minute call is not received from the working team, the rescue service is called.[23]

REFERENCES

1. Institution of Civil Engineers. *Civil Engineering Procedure* (1979)
2. P.M. Hillebrandt. *Analysis of the British Construction Industry.* Macmillan (1984)
3. D. Barber. *The Practice of Personnel Management.* Institute of Personnel Management (1978)
4. B. Fryer. *The Practice of Construction Management.* Collins (1985)
5. R. Fellows, D. Langford, R. Newcombe and S. Urry. *Construction Management in Practice.* Construction Press (1983)
6. R.E. Calvert. *Introduction to Building Management.* Butterworths (1981)
7. R.H. Clarke. *Site Supervision.* Telford (1984)
8. G. Peters. *Project Management and Construction Control.* Construction Press (1981)
9. D. Lock. *Project Management.* Gower (1984)
10. J.F. Woodward. *Quantitative Methods in Construction Management and Design.* Macmillan (1975)
11. A. Lester. *Project Planning and Control.* Butterworths (1982)
12. J.A. Milne. *Tendering and Estimating Procedures.* Godwin (1980)
13. I.H. Seeley. *Quantity Surveying Practice.* Macmillan (1984)
14. R. Oxley. Incentives in the construction industry — effects on earnings and cost. *Chartered Institute of Building Site Management Information Service No. 74.* (Summer 1978)
15. E.R. Skoyles. Waste and the estimator. *Chartered Institute of Building Technical Information Service No. 15.* (1982)
16. R. Holmes. *Introduction to Civil Engineering Construction.* College of Estate Management (1975)
17. J.R. Illingworth. *Site Handling Equipment.* Telford (1982)
18. Institution of Civil Engineering Surveyors. *The Surveyors Guide to Civil Engineering Plant* (1984)
19. P.C. Horner. *Earthworks.* Telford (1981)
20. J.M. Neil. *Construction Cost Estimating for Project Control.* Prentice-Hall (1982)
21. British Standards Institution. *BS 6164. Code of Practice for Safety in Tunnelling in the Construction Industry* (1982)
22. T.M. Megaw. *Tunnelling.* Telford (1982)
23. A. Wignall and P.S. Kendrick. *Roadwork: Theory and Practice.* Heinemann (1981)

5 Site Supervision

This chapter covers the setting out of civil engineering works, the method of supervision of work on site, the means of implementing the specification, inspecting and checking materials and workmanship and ordering the replacement of defective materials or construction, and testing procedures.

SETTING OUT

General Obligations

Under clause 17 of the ICE Conditions of Contract, the contractor is responsible for the true and proper setting out of the works. He is responsible for any errors that may arise and shall rectify them at his own expense unless the errors result from incorrect data supplied by the engineer or his representative. The checking of any setting out by the engineer or his representative will not relieve the contractor of his responsibilities, and he should, in his own interest, carefully maintain all bench marks, sight rails, pegs and other items in setting out.

Right Angles

When setting out buildings, roads and other works, it is frequently necessary to set out one line at right angles to another. The five principal methods of doing this are now outlined.

(1) The use of a builder's or timber square in the form of a large right-angled triangle as illustrated in figure 5.1. One side of the square is laid along the base line and the other side extended with a line or tape will produce a line at right angles to the first. The timber square is cumbersome but reasonably effective.

(2) Using an optical site square which is produced commercially as a site tool. It has a fixed 90° angle between two small telescopes each fitted with a 'V' sight. It is reasonably accurate but is restricted to two basic functions and has only low optical power.[1]

(3) Use of a surveyor's level fitted with a 360° horizontal circle and possibly a vernier attachment. When provided with a powerful telescope it is accurate but suffers from the deficiency of having its telescope fixed in a horizontal plane which causes problems on undulating sites.

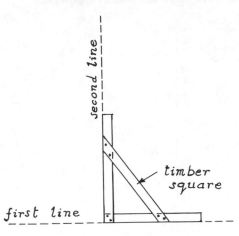

Figure 5.1   Timber square

(4) The ideal but most expensive and complex instrument is the theodolite in which the telescope can be rotated in both horizontal and vertical planes. Two graduated circular scales are also attached to the instrument, one in a vertical and the other in a horizontal plane, enabling maximum flexibility of use.

(5) The use of the 3:4:5 rule, derived from the thereom of Pythagoras, is illustrated in figure 5.2, whereby a steel or linen tape is laid out along the base line AB and extended along BC and back along CA. The lengths of the sides must be in the ratio of 3:4:5; AB could conveniently be 3 metres, BC: 4 metres and CA: 5 metres.

The normal procedure is for three persons to hold a ranging rod at each of the three points, and to ensure that the correct readings are obtained on the tape at these points, namely 3 m, 7 m and 12 m respectively. The ∠ CBA is then a right angle and BC can be extended to produce the required line. It can with more difficulty be carried out by a single person and is a relatively simple and effective method of setting out a right angle.

Buildings

The main front wall of a building is often set out from the adjoining buildings or other fixed points, such as the boundary of the site, to the dimensions shown on the site plan. Profiles are established at the corners of the building and wall intersections, and right angles set out often using a builder's or timber square.

The profiles consist of horizontal boards running parallel to the wall and supported by short lengths of 50 × 50 mm post, as illustrated in figure 5.3. Saw cuts are formed on the tops of the profile boards, which are often about 25 × 150 mm, and the cuts represent the edges of the walls and concrete foundations.

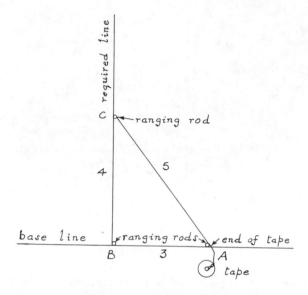

Figure 5.2   3:4:5 rule

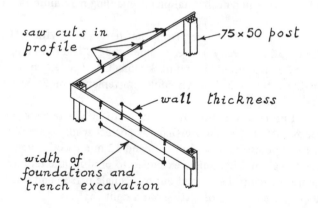

Figure 5.3   Profiles

Lines are strung between the saw cuts to which the concretor and bricklayer will work.

The setting out of a steel framed building requires extreme accuracy as the stanchions and beams are cut to length at the fabricator's works, and any error in setting out can involve expensive alterations on site. It is usual to erect continuous profiles around the outside of the building and to set out the column spacings along them. Wires stretched between the profiles will give the centres of stanchions. Lines are plumbed down at the intersections of wires to give the centre of each stanchion foundation, around which a templet is constructed. As an additional precaution, it is often considered good practice to bolt in position the beams on two floors of a multi-storeyed building prior to finally grouting in the stanchion bases.[2]

General Excavation

Excavation to formation levels preparatory to the construction of roads, structures and site works generally, is normally established by the use of profiles and a traveller. As described previously, the profile is a sight rail consisting of a timber cross board supported by timber posts. The dimensions of the sight rail depend on the durability required and the distance over which sighting between profiles is to take place. A sight rail will be set at a selected height above the finished construction level or the final formation level. Sight rails should permit easy sighting and are usually fixed at between 0.75 m and 1.80 m above existing ground level.

The traveller, which is portable and similar in configuration to a sight rail, is used to determine the depth of excavation and this is achieved when the top of its cross board coincides with the imaginary sight line connecting the profile sight rails. During bulk excavation, which may be of considerable depth, the height of the traveller must be kept to a manageable length to ease the task of transporting it over the excavated area. Hence, to keep the traveller length to a maximum of 4 m, it may be necessary to profile the excavation at various levels of dig, stepping the traveller as excavation progresses.

Alternatively, where the existing ground is well above final excavation level, digging may commence without the use of sight rails. Profiles are erected when checks with the engineer's level show the excavation to be within 1 m of the final dig level.[3]

Slope rails are used in conjunction with travellers to control side slopes or batters to embankments and cuttings. They are sometimes referred to as batter-boards. In the case of an embankment, the slope rails represent a plane parallel to, but positioned above the proposed embankment slope, to prevent them becoming covered during the earthmoving. With cuttings the posts supporting the slope rail are usually sited beyond the edge of the proposed cutting to avoid disturbance during the excavation work.

The length of a traveller from its upper edge to the base should be a convenient dimension to the nearest half metre, and will equate with the height of the sight rails above the required plane. The traveller is sighted in between the site rails and is used to monitor the cut or fill. It rises or falls according to the filling or excavation operation. Excavation or compaction ceases when the tops of the sight rails and the travellers are in alignment.[4]

Roads

Road lines are usually set out from dimensions entered on setting out drawings prepared by the engineer, on which road lengths, radii and angles of intersections are shown. Some engineers set out roads on their centre lines but the more usual practice is to set out road lines at a fixed distance back from the face of the kerbs, such as 600 or 900 mm. With the latter approach, the pegs are less likely to be disturbed when earthmoving takes place. The pegs are usually about 50 × 50 mm in section and 750 mm long, and it is advisable to surround them with concrete for protection and to paint the tops of the pegs a distinctive colour, often red, for ease of identification. The pegs are usually spaced at 30 m intervals on straight lengths and 6 m apart on curves.

Level pegs are generally positioned at the same intervals, suitably protected and painted a distinctive colour, such as blue. Level pegs are normally provided on vertical curves at about 5 or 6 m intervals and level points between the 6 m levels are normally eyed in with line and steel pins, set alongside the face of the kerbs at about 2 m intervals, to eliminate kinks in the kerb lines.

Angles are almost invariably set out with a theodolite and the base lines will be located with measurements taken with a steel tape from fixed points on the site, the necessary dimensions being read, where possible, or scaled from the drawings. *Horizontal curves* are best set out from a tangent point using a theodolite, although in the absence of an instrument the offset method can be used. The following example illustrates the normal method of setting out a curve with a theodolite, by reference to figure 5.4.

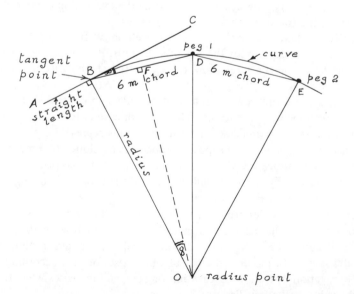

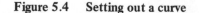

Figure 5.4   Setting out a curve

The theodolite is set up over the tangent point B and sighted back along the setting out line to A. The instrument is then transited to sight along the line to C. The first point on the curve, peg 1, is now positioned by turning the telescope through the tangential angle ($\angle$CBD) which is equal to half the angle at the centre of the circle subtended by the chord BD. The chords are usually taken in lengths of 6 m. The second peg will be positioned by setting the telescope at twice the tangential angle from B and measuring a further 6 m from D to E. The tangential angle in minutes = 1719 $\times$ chord/radius.

This process is continued until all the points along the line of the curve, at 6 m intervals, have been established.

In the offset method, the engineer measures 6 m along the line BC and then sets off a right-angled offset towards D equal to chord^2/(2 $\times$ radius). Peg 2 is then positioned by measuring a further 6 m along line BD produced and another off-set set out towards E equivalent to chord^2/radius, and this procedure continues until all the points on the curve have been fixed.

Vertical curves are incorporated where a sharp change of gradient occurs and the relevant data is normally provided on a longitudinal section drawing. The engineer in designing the road will have considered the following factors:

(1) whether a vertical curve is necessary;
(2) sight line requirements; and
(3) principal levels and gradients.

The site engineer is normally provided with calculated principal (tangent) levels and gradients and he will then calculate the intermediate levels, tabulate them and subsequently relate the designer's data to site operations.[1]

The curves are normally of parabolic form and are based on the formula:

$$X = \frac{a + b}{200 L} Y^2$$

where X = deduction from the first gradient extended
      a = first gradient
      b = second gradient
      L = length of curve
      Y = distance along curve to level point.

The following example will serve to illustrate the method of calculating the levels on the vertical curve, in this instance a summit curve as distinct from a sag curve, shown in figure 5.5.

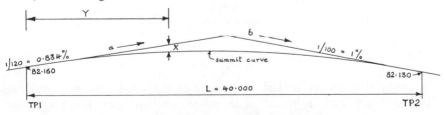

Figure 5.5   Diagram of summit curve

Applying the formula to the road profile illustrated in figure 5.5:

$$X = \frac{a+b}{200\,L}\,Y^2$$

$$= \frac{0.834 + 1.000}{200 \times 40}\,Y^2$$

$$= 0.000\,229\,Y^2$$

Thus where $Y = 5$ m, then $X = 0.000\,229 \times 25$

$$= 0.005\,73 \text{ (rounded off to 0.006)}$$

The height of tangent is obtained from $Y/120$, for example $5/120 = 0.042$, for 5 m from the first tangent point (TP1). Table 5.1 shows the method of calculating the levels at the centre line of the road at 5 m intervals along the vertical curve, with all dimensions in metres.

Table 5.1 Summit curve table

Distance along curve (Y)	5	10	15	20	25	30	35	40
Height at first tangent point (TP1)	82.160	82.160	82.160	82.160	82.160	82.160	82.160	82.160
Additional height of tangent extended	+0.042	+0.084	+0.126	+0.168	+0.210	+0.252	+0.294	+0.336
Deduction from tangent extended (X)	−0.006	−0.023	−0.052	−0.092	−0.143	−0.206	−0.281	−0.366
Height of curve (AOD)	82.196	82.221	82.234	82.236	82.227	82.206	82.173	82.130

Sewers

To ensure that sewers are laid to the correct levels and gradients, sight rails consisting of horizontal boards, often painted alternately black and white and supported by posts, are set up at all manholes. Sometimes the posts are supported by gravel rammed into pipes placed vertically. The sight rails are usually positioned with the aid of a dumpy or precise level at a convenient height above the invert level of the sewer for sighting purposes. Intermediate invert levels of pipes are determined with boning rods (travellers) sighted between the sight rails as shown in figure 5.6.

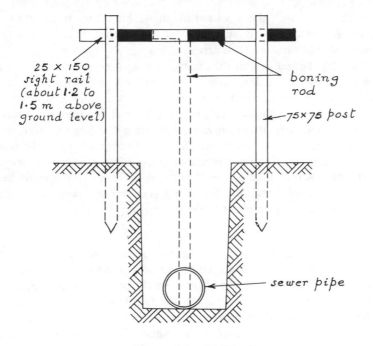

25 × 150 sight rail (about 1·2 to 1·5 m above ground level)

boning rod

75×75 post

sewer pipe

Figure 5.6   Sight rails

Tunnel Work

The setting out of tunnel work is an extremely complex and precise operation, and a method of setting out sewers to be constructed in tunnel is now described. The need for extreme accuracy will be appreciated, when it is realised that the driving of tunnels and sinking of working shafts will be proceeding simultaneously.

The sewer lines are first set out above ground using traverse methods, preferably with a micrometer transit theodolite incorporating a scale with 10 second divisions and a 30 metre steel band. A large number of readings are needed of each angle where there is a change of direction of the line of sewer. Reference points are established on two sides of each manhole beyond the excavation limits, usually consisting of bolts set in concrete on which the exact lines are marked with file cuts.

To commence the tunnelling work, the first working shaft is excavated, usually in the middle of the project where work is to proceed simultaneously on two tunnel faces. Wires are dropped from fixed points above ground to give the base line of the sewer below ground and this line can be extended by one of two methods.

(1) *The Co-planer Method* in which the instrument is set up in a pilot heading in perfect alignment with the wires and the theodolite is then transited to produce the line. The main disadvantage 'of this method is the difficulty of setting up the instrument in this precise location.

(2) *The Weisbach Method* in which the instrument is set up slightly off line, and the angles between the wires and points fixed on provisional lines in the headings are taken, from which, after a series of calculations, it is possible to determine the extent of any errors in alignment and the minute corrections that will be necessary at the provisional points in the headings to give the precisely correct alignment. Further checks will be carried out as the tunnel is driven in each direction and corrections made when necessary. The lengthy calculations involved in the second method result in this process being more time consuming than the first, but it is the more accurate and cancels out any errors in the adjustment of the instrument.

When the correct line has been established, line dogs (metal brackets) are bolted to the crown of the segments which normally make up a large diameter sewer. File cuts will be made on the dogs to give the precise line and string lines are dropped from the two line dogs nearest the face in each drive, to guide the miners who normally use these lines to align a piece of white card on tramel sticks held horizontally at the axis of the tunnel face. All the line dogs are suitably referenced.

Hedley and Garrett[5] have described the following procedure for setting out sewers in tunnel using lasers.

(1) Construct a suitable base such as 8 pipes or 20 tunnel rings, using a theodolite or level.
(2) Position laser unit over a predetermined point, usually on the centre line of the tunnel, and connect to a 12 volt DC battery.
(3) Set the beam of the laser unit at a prescribed distance above or below a given datum, usually a pipe invert or tunnel soffit.
(4) Level the laser unit both horizontally and vertically.
(5) Set the laser unit to the required alignment using the established baseline.
(6) Programme the laser unit for the required gradient.
(7) Set the target centre at the same distance above or below the datum in (3).
(8) Carry out frequent checks to ensure accuracy.

METHOD OF SUPERVISION OF WORK ON SITE

General Organisation

The resident engineer's site team has the primary function of supervising the contractor's activities, and, as described by Clarke[6] the team must to a large extent provide its own driving force. Admittedly there are many occasions when the contractor generates action as, for example, by requesting approval to place concrete or to fill pipe trenches. Nevertheless, the most routine supervision is undertaken by the engineer's site staff acting on their own initiative.

The resident engineer can establish procedures for regular inspections and encourage his staff to maintain a close watch on the contractor's operations, but much will depend on the enthusiasm and diligence of the team. The individuals making up the team work together as a closely knit group. The resident engineer has to ensure that they are provided with a satisfactory working environment to achieve the best results.[6]

Figure 5.7 shows an organisational chart covering the engineer's site team employed on a contract for a motorway. In this example, the resident engineer is backed up by three assistant resident engineers each having 'line management' responsibilities for a major part of the project. The appointment of a deputy resident engineer is rarely justified.

The main responsibility of assistant resident engineers is to manage the team's day to day activities. An assistant resident engineer ensures that staff are effectively deployed, adequate records maintained and all necessary measurements agreed and recorded. He oversees routine communications with the contractor, liaises with statutory undertakers and other third parties, supervises any new design work and modifications to the contract drawings and controls the arrangements for approval of work on site.

The section engineers report to the assistant resident engineers, and their duties may be allocated on a locational or functional basis, depending on the nature and phasing of the works. The section engineers form the grass roots end of site supervision. They should be fully aware of what is taking place on the site within their particular sphere of interest and should ensure that adequate records are kept and matters of significance reported to the assistant resident engineer. They are especially responsible for the effective operation of the processes of inspection and supervision.

Each section engineer normally leads a group of personnel made up principally of technicians and inspectors. Engineers and technicians concentrate on such activities as checking calculations and setting out, carrying out survey work and assessing performance, while inspectors, who are normally recruited from trade operatives, are mainly involved with the supervision of workmanship and the method aspects of the specification.[6]

It is customary to find some specialists among the engineer's site staff such as a quantity surveyor and materials engineer. The quantity surveyor is concerned

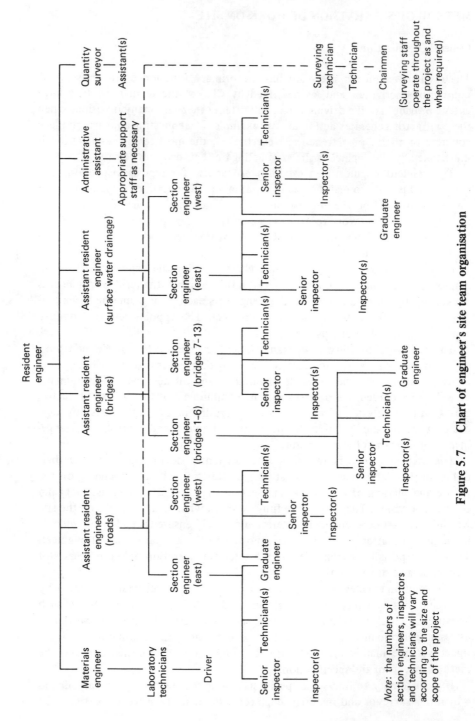

Figure 5.7   Chart of engineer's site team organisation

Note: the numbers of section engineers, inspectors and technicians will vary according to the size and scope of the project

with the measurement and valuation of the contractor's work and is sometimes referred to as a measurement engineer on a civil engineering contract. The materials engineer will be involved in the testing of materials and components and reports on the tests and their implications.

Apart from the involvement of the engineer's site staff with the inspection of work and materials to ensure compliance with the specification, the contractor will also institute his own quality control procedures, in addition to comparing progress with the approved programme. Most of the contractor's supervisory work will be carried out by a general foreman usually responsible to a resident agent and project manager. The position with regard to labour, plant, materials and sub-contractors will be continually monitored. The contractor is required under clause 15(2) of the ICE Conditions of Contract[7] to have a competent and authorised agent or representative, approved in writing by the engineer, constantly on the works and giving his whole time to their superintendence.

Resident Engineer's Duties

The resident engineer should ideally follow a prescribed procedure for supervision and inspection. This procedure should define the matters to be inspected, the frequency of inspections and the persons responsible for the inspections. It is important that the correct materials and an acceptable quality of workmanship should be secured. The origin and source of supply of materials should be checked, samples obtained and tested where necessary, manufacturer's test certificates obtained and examined and the contractor's storage and handling facilities on site approved.[8]

With regard to workmanship, tests such as compaction tests on filling, cube tests on concrete and checking of the film thickness of protective coatings are carried out regularly and the results recorded. Good workmanship is dependent on the use of suitable materials, the quality of craft operatives, suitability of plant and tools, and satisfactory arrangements for carrying out the work, including formwork, stagings and protective screens and coverings.[8]

Some specifications prescribe higher standards than the project requires. This is unfortunate as both the contractor and the engineer's site staff may form the view that specifications do not always mean what they say and that lower standards may be accepted. The specification is a contract document which should be interpreted reasonably by all parties to the contract.[9]

The resident engineer should not threaten the contractor unless he is able to implement his threats and where he is unable to make any impact on the contractor by persuasion. Ballantyne[9] has identified three alternative approaches depending on the authority delegated to him by the engineer:

(1) to use to the utmost the disciplinary powers delegated to him;
(2) to threaten to report to the engineer; and
(3) to threaten to report to the employer.

None of these actions should be taken unless the resident engineer has first confirmed that the engineer or employer will support him.

The employer's main sanction is to employ another contractor to remedy defective work and subsequently to recover the cost from the defaulting contractor, under clause 39(2) of the ICE Conditions,[7] or he may remove the contractor's name from the list of selected contractors.

On occasions the contractor does not allow the resident engineer sufficient time to inspect materials or workmanship. The ICE Conditions of Contract[7] provide for the supply of material samples (clause 36) and for the contractor not to cover up any work before it has been approved and the measurements agreed (clause 38). It is advisable to remind the contractor of these provisions before work is started in order that he shall allow adequate time in his programme for inspections and measurements.

Practical Examples

Some examples are now given illustrating the general approach to the supervision of specific types of work.

Roadworks

On a motorway contract the employer's site engineers will undertake a variety of duties which are likely to include the following activities:

(1) organising the collection and processing of data on the output of earth-moving plant;
(2) making arrangements for temporary traffic signals and major diversions;
(3) carrying out earthworks measurements for monthly certificates often in conjunction with a site quantity surveyor;
(4) carrying out inspections of newly-erected fencing with adjacent landowners; and
(5) undertaking overall control of the site supervision of contractor's work and agreeing any remedial work required.

Technicians working under the direction of an engineer usually carry out routine levelling, record the position of soil samples taken by the materials engineer, accompany the photographer on his monthly visits, make checks on the contractor's setting out for earthworks and drainage, and check the lines and levels of completed pipework.

Senior inspectors normally record the deployment of plant and labour, make random checks on round-trip times for motorscrapers and their average loads, check the condition of haul routes and road crossings, assess, with a section engineer, the extent of unsuitable material exposed in the main cutting, and general supervision of earthmoving. They are supported by inspectors who carry out the detailed supervision of excavation operations and drainage work.[6]

Pipework

The following list represents a selection of the more important items that require checking on pipework installations.

(1) pipe barrels, seams and ends for signs of damage;
(2) internal linings and external protective coatings for signs of damage;
(3) ensure that all pipe joints are effective;
(4) pipe supports and thrust blocks to ensure that the correct type is used;
(5) position of all fittings and specials, such as T-pieces, Y-pieces and reducers;
(6) positions of and supports to valves;
(7) ensure that all temporary restraints, spacing pieces and supports are removed on completion of the installation;
(8) height and size of plinths at duckfoot bends and whether the bends need to be temporarily supported during installation;
(9) spacing and fixing of brackets supporting vertical pipework;
(10) ensure that the correct colour coding and means of identification have been used for pipes and fittings, as appropriate;
(11) specified pressure tests on complete installations or sections, as required;
(12) paintwork on pipework supports; and
(13) satisfactory detailing where pipes pass through walls and slabs.[10]

Pump and Motor Sets

Typical items that require checking include the following:

(1) service connections to the motor, such as fuel, cooling water, electricity supply, exhaust controls and instrumentation;
(2) suction and delivery pipework connections and valves and supports for them;
(3) level of suction inlet;
(4) levels of control electrodes;
(5) heights of plinths and bedding and fixing of base plates;
(6) floor drainage;
(7) ventilation of pump room; and
(8) ensure that the set is the correct one for the particular location.[10]

Temporary Works

Adequate supervision of temporary as well as permanent works is required and some examples are given to illustrate this aspect.

Scaffolding   Before a scaffold is used it should be checked to ensure that:

(1) It is in accordance with the design and made up of satisfactory components.

(2) It is satisfactory for the purpose which it is to serve. For example, the decks must be at the required levels and suitably placed to provide access to the constructional work, and be free from traps, such as awkwardly placed openings. The actual loads carried should not exceed those incorporated in the design.

(3) It is assembled properly with the components positioned accurately and all bolts satisfactorily tightened.

(4) A responsible supervisor has signed the register.[11]

Falsework Falsework is a temporary structure which enables the permanent works to be constructed, and which is retained until the permanent structure is self-supporting. The extensive use of reinforced concrete has extended the need for falsework, and many modern structures of timber, steel and brickwork also require temporary support during their construction.

Falsework can be designed from first principles, but as many of the problems are repetitive, standard solutions are often applied. For example, data relating to common slab thicknesses and beam sizes, using props or scaffold, and timber runners and bearers is given in the Falsework Code of Practice.[12] This constitutes a convenient method of determining falsework requirements where problems are relatively straightforward and design facilities are lacking, but it may not produce the most economical solution.

It is essential that all falsework on a civil engineering contract shall be thoroughly checked. If the falsework contains substandard or incorrect components, it may be dangerous. If it is erected carelessly or deviates from the designer's intentions, it will not behave as planned. Furthermore, it is important to check that the conditions of use of the falsework have not changed from those envisaged at the design stage.[13]

The erected falsework should be checked against the final design drawings and any other supporting documentation. The quantity and quality of materials and components must satisfy the design requirements.

The thickness and weight of standard rolled steel sections can be determined by measurement and reference to steelwork tables, but the quality can be established only by a laboratory. Similarly, timber sizes can be readily checked but species and quality or grading are more difficult to assess, although the use of BS 5268[14] does simplify the task.

The method of assembling the various components is often critical. Relatively small inaccuracies in setting out can result in substantial changes in the way in which the falsework carries the loads. Before loading, a falsework co-ordinator should advisably sign a permit to load. He will do this after completing a satisfactory inspection. The permit should state clearly the extent of the approval in terms of the area and load to be applied.[13]

The falsework must not be dismantled before the permanent structure is able to support itself. The exact requirement may be prescribed in the specification, or the permanent works designer may issue instructions. Alternatively, it may be necessary to calculate or assess when dismantling is feasible. The co-ordinator should issue a permit to dismantle.[13]

Headings Headings are small tunnels cut into the sides of trenches or shafts and they constitute the most hazardous section of temporary work. The cutting of headings requires very careful investigation and consideration in relation to design, quality of materials, excavation techniques and workmanship, in addition to the safety and supervision implications.[15]

The routine inspection must be carried out at the advancing face of the excavation, which constitutes the point of maximum risk. The speed of the excavation has an important bearing on safety and the rapid erection of a support structure at the cut face is critical. In consequence, inspection and supervision must be carried out by personnel who are experienced in this class of work and who can inspect, approve and authorise continuation of the work without retarding the speed of advance on which the safety of the work may depend.[15]

The Timber Research and Development Association[15] has prescribed minimum sizes and grades of timber for use in headings not exceeding 2 m in height or width. Support for larger headings must be designed by a competent engineer.

IMPLEMENTATION OF THE SPECIFICATION

Supervisory Implications

Clarke[6] has aptly identified four principal supervisory functions of the engineer, all of which have direct relevance to the specification.

(1) *Precautionary measures*

The engineer's site staff can, with advantage, use their knowledge of the site and the design to identify possible problem areas and to alert the contractor to them and to suggest methods by which the problems can be overcome. The objective is to assist in the smooth running of the contract by avoiding the execution of abortive work. It must however be appreciated that the contractor has sole responsibility for carrying out the works and for deciding the methods to be used.

(2) *Overseeing constructional work*

The engineer's site staff and, in particular, the inspectors keep a close watch on the standard of workmanship, having regard to the specification requirements and generally accepted standards of good practice. Some specification requirements are more easily enforced than others. For example, the curing of a soil cement base by spraying with 55 per cent bitumen emulsion at the rate of 0.75 litre/m^2 after compaction can be readily checked, while the adequacy of compaction is more difficult to establish. A typical specification clause requires that within 2 hours of mixing, a 2 to 3 tonne smooth-wheeled roller shall be used followed by a 8 to 10 tonne smooth-wheeled roller.[16] It is also possible that equally satisfactory results can be achieved by using alternative methods to those prescribed in the specification. Any deviations from the specification require prior approval and careful scrutiny by the engineer.

Supervision of each and every activity on a large site to ensure compliance with the specification may prove very difficult or even impossible in practice because of the limitations on the resources available. Furthermore, the contractor is not obliged to notify the resident engineer of the location and timing of every individual operation. Although he is required under clause 38 of the ICE Conditions[7] to give notice of any work which is to be covered up or put out of view.

However, there are usually provisions in the contract documents for the submission of weekly programmes and for the giving of advance notice of important activities such as concrete pours and operations to be undertaken outside normal hours. The extent of supervision also varies with the type of work. For instance, compaction of extensive areas of fill, carriageway surfacing and structural concrete pours all require full-time supervision, while other operations, such as pipe laying, manhole construction, formation of slopes to cuttings and embankments and placing of steel reinforcement, can be covered by periodic visits. It helps if the contractor is also making his own thorough supervisory arrangements.

(3) *Checking of performance standards*

The engineer's site staff are actively involved in checking that the work complies with the performance standards prescribed in the specification. The engineer's supervisory staff will themselves carry out many checks, such as surface tolerances on road construction, asphalt finishes, and concrete faces after striking of formwork. Materials and components will be inspected on delivery to the site and samples of completed work sent to the laboratory for testing where appropriate. Pipe runs must be checked for satisfactory lines, levels and joints before the trenches are backfilled, and formwork and reinforcement approved before concrete is placed, all involving urgent action at the right time. Intermediate checks include a visual inspection of concrete as it leaves the truck-mixer, looking for the British Standard kite mark on products conforming to British Standards and inspection of test certificates accompanying pipes.

(4) *Approving and monitoring contractor's proposals*

The contractor will, in the early part of the project, submit to the engineer for his approval, details of his methods of working, sources of materials and suppliers of components. In some cases the engineer will require the construction of sample panels of materials, such as bricks and concrete finishes, for approval prior to starting the work on the project. In other cases the engineer will wish to inspect the works where products are produced, such as gravel quarries and fabricating yards. Such visits do not, however, preclude the engineer from subsequently rejecting the products on delivery to the site as failing to conform to the specification. The contractor is responsible for ensuring that all workmanship and materials satisfy the requirements of the contract.

Compliance with the Specification

The engineer's site staff have continually to make judgements as to whether the contractor's methods, materials and workmanship comply with the specification, or in the absence of a detailed statement of performance, accord with the normally accepted standards of good practice. Clarke[6] has described six different approaches depending on how the requirements have been expressed.

(1) The specification may be extremely rigid in its requirements. For example, the hooked ends of bar reinforcement shall each have an internal diameter of curvature and a straight length beyond the semicircle of at least four times the diameter of the bar. The decision as to compliance or non-compliance is determined by a direct comparison with the prescribed standard.

(2) Some specification provisions are in the 'deemed to satisfy' category, such as where the supplier of a specific material or component is listed which effectively pre-determines compliance and eliminates the need for individual assessment. However, where the material is found to be substandard on delivery to the site, the engineer can reject it.

(3) Certain tolerances may be permitted such as a compacting factor of 0.82 to 0.85 for vibrated concrete in reinforced concrete floors, beams and slabs. This establishes a range of compliance against which individual results are assessed. Other examples include bricks and precast concrete units. It would be unwise for the engineer to accept small tolerances in component dimensions outside the limits.

(4) Where large quantities of components are involved such as a structural concrete, it is customary to work to a statistically based standard of compliance, thus determining the degree of consistency of the end product instead of concentrating on individual samples. Adjustments can then be made to ensure the achievement of satisfactory standards.

(5) A specification requirement may be accompanied by a discretionary power of acceptance of an alternative based on the supervisor's own judgement. Thus pile shoes could be specified as chilled hardened cast iron with mild steel straps weighing 25 kg each, as supplied by Messrs X or other equal and approved. This approach can sometimes cause problems as the engineer may not always approve the alternative product which the contractor selected and allowed for in his tender.

(6) A specification may include some items of work that are done 'as required by the engineer' or 'to the satisfaction of the engineer'. For example, the contractor may be required to keep all excavations free from water from any cause by pumping, drainage or other means to the satisfaction of the

engineer and for such period as he may require. Compliance then becomes a matter for the supervisor's discretion, subject only to the customary standards and practice of the construction industry.

Requests by the contractor for a relaxation of a specification requirement should normally be rejected as to do so constitutes a departure from the contract. In exceptional cases the resident engineer may be prepared to investigate the matter in conjunction with his colleagues and possibly examine the use of the substitute item on other projects.

Clarke[6] has described how circumstances may arise where a proposal by the contractor meets the requirements of the specification but does not produce the optimum result. Supervision is concerned with two criteria — adequacy of the permanent works and compliance with the specification, and a conflict can arise between the two on occasions. Hence compliance with a particular requirement of a specification will be a necessary condition for the acceptance of a contractor's proposal but it may not be a sufficient condition.

Problems in Implementing Specifications

Problems sometimes arise in the implementation of specifications because of problems of interpretation. The document may, for example, contain excessively long paragraphs consisting of wording which is difficult to understand. The fault may lie in attempting to describe in a specification details which are better depicted on a drawing. What may be clear to the specification writer can be ambiguous to the contractor's staff and particularly to site operatives. The best specifications are clear, concise and precise, and long specifications often contain many irrelevancies.

One of the greatest dangers in specification writing is to work from an old specification without giving sufficient thought to the project in hand. When a contractor observes a number of inappropriate clauses, he may take the view that some of the relevant clauses also do not apply. A few engineers who have had the unfortunate experience of omitting important provisions from a specification have subsequently inserted every conceivable item in future specifications, supposedly to safeguard their interests, thus causing contractors considerable additional work in pricing the projects and in executing them. Another unsatisfactory approach is to insert 'escape' clauses in the specification, whereby the contractor is responsible for carrying out all the work which is omitted from the specification, but which is necessary to complete the permanent works. This approach can result in the submission of inflated tenders and acts against the employer's best interests.[17]

A specification should be complete, authoritative and up-to-date. It should be relevant to the specific project and readily understood. Great care should be taken when using an office or type specification as a guide to ensure that it contains no irrelevant clauses, covers all the features of the project in hand and makes reference to all the latest relevant British Standards and Codes of Practice. Great difficulty can be experienced by the engineer's site staff in attempting to implement an unsatisfactory specification which will inevitably lead to unnecessary and time consuming disputes.

MATERIALS REQUIREMENTS

Procedural Aspects

The main procedural aspects relating to the supply of materials and components on site have been described by Elsby[18] and are now listed.

(1) The resident engineer should obtain and keep for reference all relevant British Standards and Codes of Practice.
(2) He should arrange with the contractor the procedure for reviewing and approving proposed suppliers of materials and components for permanent works.
(3) A check should be made of the prices of materials and components included in any basic rates schedule.
(4) The orders for materials and components should be checked to ensure that they comply with the contract.
(5) A register of orders should be maintained, incorporating order numbers, dates, suppliers and sufficient information to identify the materials and components being supplied. The dates of inspection on and off the site should be recorded, together with dates of tests and reference made to test certificates.

All materials should be obtained from approved manufacturers or suppliers and comply with the requirements of the current appropriate British Standards. A list of suppliers should be submitted by the contractor as soon as possible after the award of the contract. Suppliers must be willing to admit the engineer or his representative to their premises at all reasonable times to obtain samples or check the process of manufacture. Suppliers must not be changed without the prior approval of the engineer. Samples shall be supplied without charge and be delivered by the contractor to wherever directed. Approved samples may be retained by the engineer's site staff as a standard for comparison of subsequent supplies. Manufacturers' test certificates will normally be accepted as proof of compliance with the test requirements of the appropriate British Standards. The contractor is responsible for collecting or taking delivery of all materials and components for use on the works and for arranging suitable storage.[5]

Concreting Materials

Most *cement* used on large civil engineering projects is stored in silos which are usually filled from 20 tonne bulk tanker trailers so they need to hold a minimum of about 35 tonnes to ensure continuity of concrete production. Bags of cement may be used on the site to make up small quantities of concrete or where special cements are required. Cement delivered in bags should be stored in dry, weather-proof, well ventilated buildings of adequate size, stacked not more than six to eight bags high on pallets and used in order of delivery. Any cement which has become partially set should be rejected.[19]

Aggregates are usually delivered by tipper truck to stockpiles located adjacent to the mixer. They should be inspected before tipping to ensure that they are of the correct type and size and are clean. These subjective tests need backing up with periodic British Standard tests, as described later in the chapter. Rubbing a sample of aggregate in the hands will detect excessive silt.[19]

The aggregates should be stored on concrete bases laid to drain moisture away from the mixer. The stockpiling area should be subdivided into bays by substantial, high walls to separate the different aggregates. The most common form of construction is vertical H section steel stanchions with timber or concrete panels slotted into them. The bays should be suitably labelled to identify their contents and covered to protect them from falling leaves, rain, frost and snow.[19]

Where the contractor wishes to use *ready-mixed concrete*, he should make written application to the engineer giving details of the firm and depot from which it will be obtained. The minimum cement content should be clearly shown on each delivery ticket, so that the inspector can satisfy himself that the concrete will attain the required strength and durability. Regular checks by slump tests and test cubes are also necessary. In addition a check is required to ensure that the concrete is being delivered within the period specified in clause 7 of BS 5328,[20] namely within two hours of mixing for a truck or agitator and one hour for non-agitating plant.[21]

Steel reinforcement should comply with the appropriate British Standards and be free from oil, grease, dirt, paint and loose rust prior to use.[16] The reinforcement should be well stacked on suitable hard and well drained surfaces with adequate support to prevent the heavy steel bundles deforming under their own weight.[22]

Bricks

Site supervision will be mainly concerned with ensuring that the correct grade of bricks is being used. The bulk of bricks used on civil engineering contracts are either class A or class B engineering bricks complying with BS 3921. The bricks should be hard, sound, well burnt, uniform in texture, regular in shape, even in size and with true square arrises. Care must be taken in unloading, stacking and handling, as all chipped or damaged bricks should be rejected.[16]

Mortar for use with engineering brickwork normally consists of one part of Portland cement to three parts of sand complying with BS 1200, table 1. The sand requires approval by the engineer's site staff and must be protected from contamination. The sand can be tested by mixing with water for one minute in a glass cylindrical jar in the proportions of one part sand to two parts water. After settling for two hours, the top layer of fine materials should not exceed 5 per cent of the volume of solid matter in the jar.

Timber

The use of structural timber is governed by BS Code of Practice CP 112/BS 5268 and BS 4978. The inspector will be checking for the existence, position and extent of defects such as knots, waney edges, sloping grain, shakes and splits which reduce the strength of the timber. Machine stress grading is undertaken to measure the relative stiffness and therefore the strength of the timber under test. Any timber showing more than than 15 per cent sapwood on an end section should be rejected. All timbers must be of the scantlings shown on the drawings, with an allowance of 3 mm for each wrought face.

Steelwork

Workmanship and general fabrication procedure should be in accordance with BS 449 for buildings and BS 153 for steel girder bridges. All butting members should have their ends machined after fabrication. All members require checking to ensure that they are to the dimensions shown on the drawings, cut to exact lengths and finished true and square. All holes should be accurately marked from templates and be drilled to give smooth edges. After checking or testing at the fabricator's works, all members and fittings should, for purposes of identification during erection, have a distinguishing number and letter painted and, where possible, also stamped on in two positions.[16]

Despite their strength, structural steel components can be damaged during loading, transporting, unloading and stacking, with light section assemblies, such as roof trusses, being particularly vulnerable. Damaged items should be notified and repaired, or replaced if irreparable.[22]

Road Surfacing Materials

The most widely used materials in bituminous road construction are tarmacadam, bitumen macadam and asphalt. The first two are mixtures of graded coated stone, obtaining their strength largely from the mechanical interlocking of the aggregates. Asphalts are dense impervious mixtures of filler, graded aggregate and bitumen whose strength depends mainly on the viscosity of their bitumen binder. The nature and grade of the aggregate will require checking and also the thickness and number of courses. The testing of bituminous materials is covered later in the chapter.

Pipework

A check is needed on the types, classes and sizes of pipes to ensure that they satisfy the requirements of the contract. With the exception of plastic pipes, most varieties can be stored in the open. Plastic-based pipes degrade in sunlight and therefore require covering, they are also easily damaged and should be handled carefully. All pipes should be handled and stacked in accordance with the manufacturer's recommendations to prevent damage. In particular pipe ends should be checked for possible damage, otherwise defective joints could occur. The linings of steel pipes should also be inspected for soundness and any defective areas suitably repaired.

WORKMANSHIP REQUIREMENTS

Earthworks

It is important to check that all topsoil which is stripped and retained for subsequent use is of acceptable quality. The dimensions, levels, lines and profile of excavations require checking to ensure that they are in accordance with the contract drawings and supporting data. The formation requires approval by the engineer and any excess excavation normally has to be made good in concrete.

The engineer's site staff will check that all fill is of suitable materials, laid to the prescribed thicknesses and adequately compacted. Effective control is necessary of all earthwork support and excavated surfaces, and to ensure that all excavations are kept clear of water.

Tunnelling

All excavation must be carried out with greatest care and the methods and materials to be used require the approval of the engineer. The engineer will closely monitor the progress of the excavation to ensure that it is performed as speedily as possible and adequately supported, with particular care being taken to ensure no settlement or damage is caused to any property. The engineer will normally insist that any discontinued excavation which is exposed for more than 48 hours shall be close timbered and grouted.

The engineer will tightly control the alignment of shafts and tunnels. Commonly adopted maximum tolerances are 40 mm for ungrouted expanded linings erected behind a shield without tailplates, and 25 mm for all other linings. The engineer normally requires the contractor to provide suitable shields, and to erect, maintain and modify them as necessary, all to the engineer's satisfaction.

An inspector will check that each complete ring of bolted lining is grouted as soon as possible after erection. The space between the outside of the tunnel lining and the surrounding ground must be completely filled by cement grout.

Piling

Precast concrete bearing piles should be carefully checked for damage at all stages from stacking in the casting yard to their placing in position ready for driving. An inspector will need to check that no piles leave the yard within two weeks of casting and that no piles are driven less than six weeks after casting.

The engineer will control the type of pile-driving equipment and take steps to ensure that pile heads remain undamaged and the piles are accurately pitched and driven truly vertical. He will also decide the depth of penetration into a non-cohesive stratum and keep a continuous record of the penetration per blow as each pile is driven.

With *in situ* concrete piles the engineer's site staff will continually check the pouring of the concrete to ensure that it is adequately consolidated. It is generally provisionally assumed that the lower ends of the piles should terminate 1.5 m below the top of the bearing stratum. This depth may be adjusted by the engineer in the light of the information obtained from the borings or loading tests.

Steel sheet piling should be checked regularly during driving, to ensure that it is in correct alignment. The engineer or his representative will also check the piling equipment, that the correct type and length of pile is being used and that the piles are driven to the prescribed depths.

Concrete Work

The engineer's site staff will pay particular attention to the quality of concrete as it forms such an important part of almost every civil engineering contract. The mixer driver is a key member of the contractor's workforce as he should monitor visually the workability, homogeneity and cohesiveness of each mix, as well as the consistency of production.[19] The engineer's inspector will liaise closely with the mixer driver to ensure the supply of a consistent mix of concrete of the required quality.

One of the main aims of the inspector will be to avoid excessive segregation of the constituents of the concrete. For instance, a non-cohesive concrete mix of high workability may suffer loss of mortar as the concrete passes down a chute. Segregation may also arise when concrete is conveyed in general purpose dumpers across uneven site roads.[19]

Concrete which is being transported in open-topped vehicles, such as tipper lorries on motorway contracts or ready-mixed concrete vehicles, should be covered to avoid wetting and/or drying of the top layer of concrete. The inspector must also check the vehicles prior to use to ensure that they are not contaminated in any way. Even excess water from washing down, lying in the bottom of the vehicle, can adversely affect the quality of the concrete. Workability of the concrete decreases with time, so the longer the placing is delayed, the stiffer it will become making it more difficult to compact.[19]

All rubbish must be removed from the area which is to receive concrete. The inspector should also ensure that the concrete is deposited as quickly as possible in its final position and thoroughly compacted in the formwork and around the reinforcement, to avoid honeycombing, segregation and loss of liquid and cement. On most civil engineering contracts the concrete is compacted by vibration, accelerating the escape of gas. Internal vibrators are the most commonly used type as they are very efficient.

The engineer's site staff will closely monitor the vibration of concrete in likely problem areas. Locations requiring special attention are corners, openings, sloping slabs and large deep pours. Foundation bases may be 2 or 3 m thick, consisting of large volumes of concrete to be placed in a single operation. The inspector should ensure that the area of exposed concrete is kept to a minimum.[19]

Underwater concreting is particularly difficult and the main problems are usually overcome by using concrete of high workability and placing it by means of a tremie, which consists of a number of demountable steel tubes topped by a hopper. The tremie method can be used for tall pours, such as bridge piers and walls, and large flat pours like dock bases.[19]

The engineer's site staff will need to check frequently that newly placed concrete is protected from the harmful effects of sun, drying wind, cold and excessive water, a process known as 'curing'. In hot weather the concrete is often protected by covering it with hessian or other absorbent covering which is kept moist. In cold weather, the protective element is likely to be fibre quilts filled with plastic, glass wool or straw. Alternatively, a membrane-curing liquid may be applied. The curing period is normally a minimum of 14 days.

Concrete should not be mixed or placed when the air temperature is less than than 1°C when the temperature is rising, or less than 3°C when the temperature is falling. Concreting is best carried out continuously between construction joints, the position of which normally requires the engineer's approval. The engineer is unlikely to permit the formation of construction joints in close proximity to the junctions of walls, base slabs, roofs, columns and beams. The engineer's representative will check to ensure that construction joints, expansion joints and water bars meet the requirements of the specification. He will also check that all steel reinforcement is accurately placed and securely fixed in the positions shown on the drawings and is of the correct type and size. The prescribed minimum concrete cover to the reinforcement must be obtained in all members.

The inspector(s) will check to ensure that formwork is fixed to the required line, shape and surface and that it will not deflect under the weight of the wet concrete and any incidental loads. He will also ensure that the joints in the formwork are watertight. The internal surfaces of the formwork must be free of rubbish and dirt and be treated with a suitable coating to assist subsequent removal of the formwork. Striking of formwork should be carried out strictly in accordance with the specification requirements.

Formwork is often of timber, lined with plywood or hardboard, or of metal, to provide a good finish to the concrete. Where the concrete is to be permanently covered, rough formwork can be used.

The engineer's site staff will periodically inspect the precast concrete yard. A sound, level base slab is needed and the shutters should be durable and free from damage. Adequate curing arrangements are essential. Complete and systematic records will be kept of the casting of units and their scheduled curing and stacking times. All units should be referenced and date marked on casting.[22]

Brickwork

The engineer's site staff will monitor the construction of brickwork to ensure that it is built to the dimensions, thicknesses, heights and positions shown on the drawings. They will also check that the brickwork is built uniform, true and level, with all perpends vertical and in alignment. No brickwork should rise more than 1.25 m above adjoining work, and work in rising should be properly toothed and racked back. All bed and vertical joints should be filled solid with mortar as the bricks are laid.[16]   Checks will also be made on the inclusion of horizontal or vertical reinforcement where specified and on the position and condition of damp-proof courses, as well as on the correctness of the bricks and mortar being used. It is important that the bricks are not damaged by scaffolding or defaced by unsightly mortar droppings.[22]

Waterproofing

The engineer's site staff will check carefully on all laps, joints and other locations where asphalt and similar applications or painted coatings overlap. Corners are vulnerable, particularly the junction of the base slab and walls, which is frequently the most difficult part to inspect. The inspector(s) will check to ensure that all surfaces to be waterproofed are dry and free from dirt or loose material. The number of coats of waterproofer must also be checked.

Steelwork

All steelwork should be carefully unloaded and stacked on suitable softwood battens clear of the ground. The inspector will require the contractor to wire brush and patch prime any painted surfaces that have been damaged during transportation and unloading. He will, in conjunction with the steelwork contractor, check the position and levels of stanchion bases and that holding down bolts are satisfactorily located. The engineer or his representative will check the assembly plant, erection procedure and the dimensions, positions, levels and connections of the structural members. Where any misalignment occurs, the engineer or his representative will check the cause and if it does not result from incorrect assembly, will require the rectification or replacement of the members. No grouting of stanchion bases or beams should be carried out until the lining, levelling and plumbing of the steelwork has been satisfactorily completed.[22]

The engineer or his representative will carefully check the arrangements proposed for accommodating service pipes and cables through the steelwork and the preparation of any steelwork prior to encasing in concrete. Careful scrutiny of the jointing at the connections of steel members and of site painting is essential.

Structural Timbers

Structural timbers must be carefully handled and stacked and inspectors should closely examine the condition of the members. The locations and sizes of the various timbers should be checked together with the adequacy and effectiveness of the joints and connections. The supervision of erection will entail directing special attention to handling, bearings and stability.[22]

Roads

The engineer's site staff will check that the formation levels are thoroughly consolidated to the required levels to even surfaces finished to the required longitudinal and cross falls. Kerbs shall be firmly and evenly bedded to the correct lines and levels.

Concrete road slabs should be inspected for quality and thickness, and adequate compaction by suitable vibratory equipment. Fabric reinforcement is placed 50 mm from the finished surface and should terminate 50 mm from all expansion joints and slab edges. After compaction the finished surface is checked with a 3 m straight edge and any irregularities exceeding 3 mm shall be rectified immediately. Channels are checked to ensure that they have adequate falls to gullies. Any crazed or cracked areas of concrete must be suitably replaced.

Expansion joints are checked to ensure compliance with the specification and drawings and levels across the joints checked with a 2 m straight edge. A smooth finish should be provided adjoining kerb inlets to gullies and gully gratings.

With tarmacadam, bitumen macadam and asphalt, collectively, commonly referred to as blacktop construction, one of the most important factors is to obtain the correct temperature in order that the material can be rolled satisfactorily. Occasionally, however, cold constructional methods are used. The inspector will examine the material for type and thickness and also the surface during rolling. Any high spots, waves or signs of cracking will be noted. Successive rolling often eliminates these deficiencies but if not, the defective areas will have to be cut out and replaced. Other corrective measures that may be necessary are to grade off high spots and fill low spots. Each course or layer should be correctly laid on dry, clean surfaces to the appropriate lines and falls before the succeeding course is laid, having regard to the normal permitted tolerances.

Joints should be cut back square to sound material before laying further areas, and they should be staggered in successive vertical layers and primed with an emulsion tack coat. The type and weight of rollers and the method of working adopted should be checked against the contractor's method statement and the contract requirements. The paver or spreading machine should also be checked for working conditions and the number of lorries allocated to the work. As far as is practicable, the paver should be kept continuously on the move.

Some road projects entail the use of cement bound granular materials, where special attention should be paid to mixing, grading and cement content. The time between mixing and compacting must be checked to prevent the use of partially set material. When laid and rolled, the material should form a stable surface free from cracks. The inspector will check the thickness and ensure that newly laid material is sprayed with emulsion for curing purposes and the surface kept free of traffic for the prescribed period. Other checks will follow the procedures described for blacktop construction with regard to joints, rollers and working procedures.[22]

Sewers

The engineer's site staff will check all sewer runs for lines and levels, adequacy of falls, jointing, protection, where necessary, and connections to manholes. The pipes are normally laid from the low end with the sockets facing uphill.

It is advisable to excavate trial holes to locate existing services and so avoid problems in adjusting sewer lines at a late stage. All pipes must be firmly bedded to prevent subsequent damage to pipes. Trench widths should be kept to a minimum and backfilling of trenches carried out carefully with selected material around the pipes.

Manholes should be checked to ensure that they are constructed of the specified materials and to the correct dimensions. Particular attention should be directed to channels, benching, step irons and covers. Fair faced engineering brickwork is preferable to rendered walls built in common bricks.

Water Mains

The engineer's site staff will check on types, sizes and jointing of pipes, lines of pipes and the positions and types of sluice valves, air valves, washout valves and hydrants, and their associated chambers and nameplates. The backfilling of the pipe trenches and the permanent reinstatement of the surface justify a thorough inspection.[16]

TESTING ARRANGEMENTS

A diverse range of tests are carried out on various materials and components to ensure their satisfactory choice and operation. The following examples will serve to illustrate some of the more common tests carried out in connection with civil engineering contracts.

Soils

A useful method of measuring the penetration resistance of subsoil on the site and of obtaining soil samples from a considerable depth is the standard penetration test. A specialist contractor is usually employed for this work. Rougier[1] has described the equipment and procedure that is normally adopted.

A soils test which is often used preparatory to designing flexible pavement or road designs is the California bearing ratio (CBR) test. It consists of determining what load is required to maintain a penetration rate of 1.08 mm/minute using a circular plunger of 1935 mm^2 cross-sectional area. The required load is registered on a suitable load measuring ring and the results of tests carried out on either natural or, more often, recompacted soils can then be compared with standard test result curves to evaluate the bearing capacity of the soil.

The CBR test can be carried out either in laboratory or on site by means of *in situ* CBR apparatus which can be mounted on a rolled steel joist cantilevered from the rear of a suitable vehicle or fitted to a mobile frame. The *in situ* apparatus permits rapid test results to be available on site as work progresses.[23]

Concrete

The testing of cement is covered by BS 4550, Parts 1 to 6. The tests are undertaken by the cement manufacturer as part of the quality control procedure and certificates issued by the manufacturer.

The grading of natural aggregates is covered by BS 882. As part of the mix design process and of routine quality control and compliance testing programmes, it will be necessary to carry out both physical and chemical analyses of samples of both coarse and fine aggregates. The methods of sampling and testing are covered by BS 812, Parts 1 to 4, which includes tests for the physical, mechanical and chemical properties.[24]

The contractor should check the moisture contents of the aggregates in determining the volume of water to be added to each batch of concrete. He should keep sufficient equipment on the site for carrying out slump tests and/or compacting factor tests during each day of concreting in the manner described in BS 1881, and should keep a record of the tests for inspection by the engineer's representative. Table 5.2 gives a guide as to probable test limits.

The strength of concrete is usually determined by tests on cubes made, cured and tested in accordance with BS 1881, Parts 7 and 8. It is common practice to take six cubes from each section of the work during each half-day's concreting. If the minimum batch cube strength (average strength of batch less twice the value of the standard deviation of the results) is less than the specified minimum

Table 5.2 Slump and compacting factor test limits

Type of concrete	Slump	Compacting factor
Hand compacted concrete		
Mass concrete filling and blinding	40–65 mm	0.89–0.93
Reinforced concrete foundations, floors, beams and slabs	50–75 mm	0.91–0.94
Reinforced concrete columns and walls	65–90 mm	0.93–0.95
Vibrated concrete		
Reinforced concrete floors, beams and slabs	12–25 mm	0.82–0.85
Reinforced concrete columns and walls	25–50 mm	0.85–0.91

strengths, the concrete represented by the cubes shall be cut out and replaced with satisfactory concrete at the contractor's expense.[16]

Concrete in structures which are to withstand water under pressure is normally the subject of percolation tests. The contractor is generally required to supply concrete test slabs 125 mm in diameter and 50 mm thick, gauged in the specified proportions, to a testing laboratory, where they will be subjected to water pressure equivalent to 12 m head of water on one side of the slab for 24 hours. Should dampness appear on the other side of the slab, further tests are prepared with adjusted proportions of mix. This procedure is repeated, as necessary, until slabs are produced which satisfactorily meet the requirements of the test and the mix adopted in the last test will be used throughout this class of work.[16]

Brickwork

The principal tests carried out on brickwork are:

(1) crushing tests for bricks for designated brickwork;
(2) porosity tests on facings; and
(3) mortar tests.[21]

Piling

A load test may be carried out to check the pile design. The maximum test load should be the ultimate design load or probably more, as the object of the test is to determine the failure load and thus the actual factor of safety available. Working piles are often tested to 1.5 times their working load. Load testing may be carried out by jacking against anchors or a dead weight known as kentledge.[24]

A constant rate of penetration (CRP) test is sometimes carried out to determine the ultimate capacity of the pile. The load is applied to keep the pile head moving at a constant rate. The load is read simultaneously with deflection and overcomes the problem inherent in maintained load tests of being unable to quantify ultimate load with reasonable accuracy.[24]

Pipelines

Pipelines should normally be tested before any concrete haunch or surround is laid or backfill commenced. All tests should be carried out in daylight in the presence of the engineer's representative, using water suitably coloured with fluorescein. Any pipes showing leaks, sweating or other signs of porosity, should be condemned, replaced and retested at the contractor's expense.

All *gravitational pipelines* should be tested with water to a head of not less than 1.5 m, measured from the crown of the highest pipe under test. At no point should the pressure exceed the safe pressure specified for the pipes by the manufacturer. After allowing a short period for absorption, the vertical pipe at the head of the length under test shall be topped up and the water level observed for not less than 10 minutes. Should the water level drop by more than 25 mm, the cause should be sought and the defect remedied.

After the test, the water should be released from the stopper while a watch is kept on the vertical pipe, to check that the pipeline and the vertical pipe are unobstructed. No testing shall commence within 48 hours of the making of cement joints.

All *pressure pipelines* should be tested by filling with water suitably coloured with fluorescein and raising the pressure by injecting further water through a manually operated force-pump, fitted with an accurate pressure gauge, until the following pressures are obtained: class D pipes: 1.8 MN/m^2; class C pipes: 1.4 MN/m^2; and class B pipes: 0.9 MN/m^2. The pump is then disconnected from the pipeline and the required pressure maintained for 30 minutes.[16]

Road Materials

Samples of blacktop materials can be taken for testing from the mixing plant, at the site before laying or after laying. The Transport and Road Research Laboratory recommends that a trench should be cut across the width of the laid material, and the excavated material mixed and quartered to obtain a suitable sample for testing. Reasonably representative samples can also be obtained by cutting cores 200 mm or more in diameter at selected points in the finished material.[24]

A variety of tests may be performed which include the following:

(1) penetration of bitumen by an accurately made needle under a standard load;
(2) determination of temperatures at which different tars have the same viscosity;

(3) determination of the temperature at which a given tar or bitumen reaches a certain degree of softness (ring and ball test);
(4) determination of skid resistance; and
(5) standard sieve tests.[23]

REFERENCES

1. P.A. Rougier. *Site Engineering Practice.* Construction Press (1984)
2. I.H. Seeley. *Building Technology.* Macmillan (1986)
3. R.W. Murphy. *Site Engineering.* Construction Press (1983)
4. P. Roper. *Building Site Manual.* Northwood Books (1982)
5. G. Hedley and C. Garrett. *Practical Site Management.* Longman (1983)
6. R.H. Clarke. *Site Supervision.* Telford (1984)
7. Institution of Civil Engineers, Association of Consulting Engineers and Federation of Civil Engineering Contractors. *Conditions of Contract and Forms of Tender, Agreement and Bond for use in connection with Works of Civil Engineering Construction.* Fifth Edition (June 1973, revised January 1979)
8. J.R. Sweetapple. The role of the engineer's representative. *Supervision of Construction.* Institution of Civil Engineers Symposium, London (7-8 June 1984)
9. J.K. Ballantyne. *The Resident Engineer.* Telford (1983)
10. J.W. Watts. *The Supervision of Installation.* Batsford (1982)
11. C.J. Wilshere. *Access Scaffolding.* Telford (1981)
12. British Standards Institution. *Code of Practice for Falsework: BS 5975* (1982)
13. C.J. Wilshere. *Falsework.* Telford (1983)
14. British Standards Institution. *Code of Practice for the Structural Use of Timber. Part 2: Permissible Stress Design, Materials and Workmanship: BS 5268* (1984)
15. Timber Research and Development Association. *Timber in Excavations* (1984)
16. I.H. Seeley. *Civil Engineering Specification.* Macmillan (1976)
17. I.A. Melville and I.A. Gordon. *Professional Practice for Building Works.* Estates Gazette (1983)
18. W.L. Elsby. *The Engineer and Construction Control.* Telford (1981)
19. G. Taylor. *Concrete Site Work.* Telford (1984)
20. British Standards Institution. *Specifying Concrete, including Ready-Mixed Concrete: BS 5328* (1981)
21. Greater London Council, Department of Architecture and Civic Design. *Handbook for Clerk of Works.* Architectural Press (1983)
22. J.W. Watts. *The Supervision of Construction.* Batsford (1980)

23. A. Wignall and P.S. Kendrick. *Roadwork Theory and Practice.* Heinemann (1981)
24. J.R. Chantler, J. Glanville, A.B. Norrey and R. Forrester. *Basic Materials and Workmanship.* Telford (1984)

6 Communication

This chapter examines the importance of communication on civil engineering contracts and the various forms which it takes. The principal site documents and records are described and illustrated, together with the various types of meetings.

METHODS AND IMPORTANCE OF COMMUNICATION

Factors affecting Communication

Poor communication has often been a problem in the construction industry and this stems partly from the way in which the industry is organised. Site personnel come from different backgrounds and have varying contributions to make at a variety of levels. A large amount of information passes between them and this creates the need for a well organised and effective communication network. Even with a well established network, problems of communication can still arise because the information conveyed may be difficult to understand, inaccurate or misleading.[1]

It frequently happens that people concerned with a project are submerged under a mass of paper that they do not have time to read and hence it may be rendered largely ineffective. Regrettably site personnel cannot always obtain the information they require when they want it. Estimates may be inaccurate, drawings incomplete or out of date and specification descriptions and engineer's instructions ambiguous.

The extent of the communication problem may be influenced by the size of the firm or other organisation. Communications within small organisations are often good largely because there is extensive face-to-face contact between personnel. Larger firms tend to rely to a greater extent on written communications which provide a permanent record but can more easily lead to misunderstandings and delay in communication.

The acquisition of adequate communication skills by personnel does not have sufficient importance attached to it. Even managers are often poor communicators and may be unaware of this deficiency. Considerable benefits can be derived from the introduction of suitable training schemes aimed at improving the communication skills of employees and monitoring their performance.[1]

The Purposes of Communication

Communication serves a variety of functions, all of which are important in construction management. Fryer[1] has endeavoured to identify the more commonly encountered activities.

(1) *Information* is continually being exchanged between all persons connected with a civil engineering project. For example, a project manager explains company policy to a young engineer, an estimator shows a junior assistant how to calculate a unit rate for a certain work item and a bricklayer tells an apprentice how to point brickwork in accordance with the requirements of the specification.

Furthermore, information can pass in both directions — upwards as well as downwards. For instance, a quantity surveyor has a discussion with the project manager about a problem that he has concerning payments to a sub-contractor, while an apprentice pipelayer tells the foreman about his dissatisfaction with his travel allowance payments.

(2) *Fulfilment* of prescribed activities by appropriate personnel satisfactorily and within the required timescale is vital to the effective completion of the project. Many people are dependent on these actions in order to achieve the mutually desired goals. It is important that everyone should know what is expected of them and this creates the need for clear and implicit instructions. The basis of this form of communication is usually in the form of drawings, specifications, programmes and other supporting documents, but the managerial staff will need to ensure that all the essential information, including target dates, has been fully understood by the recipients.

For instance, a section engineer may ask a sub-contractor to increase his workforce to avoid causing disruption to the main contractor's programme. A foreman may request additional excavating equipment from the agent to accelerate progress in difficult ground conditions or further protective sheeting in the event of exceptionally bad weather. The contractor's agent may request a meeting with the resident engineer to resolve certain urgent problems which have arisen on the project.

(3) *Social relationships* can influence the smooth running of a contract. It is important that good working relationships are developed and maintained between all persons involved in a civil engineering project. The best results flow from a concerted team effort. With large organisations social aspects become an important factor and lead to increased co-operation between personnel.

(4) *Expression of individual feelings* permits an employee to make his views known on matters which he considers important. An employee may hold very strong views on the contribution that he can make to the work of the organisation and his place in the hierarchy. There is a need for informal discussions as well as the more formal organised arrangements such as site

meetings. Desirably a variety of channels of communication should be provided ranging from the opportunity for employees to have a talk with senior personnel and suggestion boxes wherein employees can deposit suggestions on any matters concerning the firm's policy and method of operation.

(5) *Changes of attitude* of employees are often required to keep abreast of technological, legal and economic developments. An employee should feel that he plays an effective role in the work of the organisation and that his efforts are appreciated by the management. The mere giving of instructions may not always secure the desired results. Employees need to be receptive to the action being proposed and this is best achieved by informal discussions at grass roots level. These discussions may be on a personal basis, or in groups, depending upon the particular circumstances. Operatives should always have some understanding of why they are carrying out a specific operation as well as how they are to do it.

(6) *Informal discussions* also have a part to play in the overall communication process. Most employees have the occasional chat with their fellow operatives on matters of common interest. Some of the topics discussed, such as the likely outcome of the local football team's next match, may well be regarded as irrelevant to the work of the organisation, but such diversions can be morale boosters and should not be unduly restricted unless they occupy an excessive amount of time. An operative who is unwilling to have a chat with his fellow workers is likely to be regarded as unsociable.

A manager should ideally show an interest in the affairs of his personnel. Thus a short discussion with an employee who returns after a serious illness, suffers a bereavement, gets married or has an addition to the family, reflects a caring organisation. The retirement of a long serving employee calls for a short speech and the presentation of a gift in appreciation of services rendered.

Organisation of Communication

Effective management control is dependent on an efficient method of conveying instructions and information and of ensuring satisfactory feedback. On a civil engineering project, many diverse interests may be involved, such as those of consultants, resident engineer's site staff, main contractor, sub-contractors, suppliers and the employer. Recognised channels of communication are needed to ensure that all the personnel concerned receive the information they require and when they need it.

A variety of communication channels may be used. For example, a line hierarchy links people making decisions with those who carry them out. Functional and lateral relationships link people in different sections, some of whom may contribute specialist expertise. Consultative arrangements between managers and operatives will facilitate the settlement of problems and improve working relationships.

The setting up of appropriate channels of communication is, however, only part of the process. They have to operate effectively by enabling the information to reach the right people at the right time and in the manner required by the recipient. This requires reliable sources of data, and provision for quick action and effective communication.

The use of microcomputers has assisted enormously in the storage and retrieval of information to meet a wide range of needs throughout the processes of design and construction of a civil engineering project.[1]

Methods of Communication

Communication methods in organisations have been categorised as lateral, upward and downward. For example, *lateral communication* may take place between people of similar status, such as section engineers, and is primarily concerned with the exchange of information. It can, however, also occur between persons having a functional relationship with one another, such as a quantity surveyor with a general foreman. Although sometimes discouraged by management, lateral communication often provides a valuable method of getting work done efficiently and quickly.

Upward communication provides a valuable source of feedback of useful information to management. The information can take various forms from submitting progress reports, making suggestions for improvements to organisation or working methods, to seeking guidance on how to deal with certain problem areas. Site personnel may, however, be more reluctant to submit information on such matters as escalating costs, failure of temporary works or the replacement of defective work, all of which could reflect adversely on their ability and performance. In unsatisfactory situations, reports from site tend to be delayed and/or distorted.

Modern management practice favours upward communication which can be promoted in a variety of ways, including participative management, joint consultation, suggestion schemes and grievance procedures. Employment legislation has compelled firms to give employees the opportunity to express their grievances and to receive a sympathetic hearing.

Downward communication consists primarily of management passing instructions and information to the personnel involved. The opportunity is often taken to keep personnel adequately informed about overall progress, general guidance and policy matters. It is beneficial for employees to know how their particular activities fit into the firm's overall objectives and plans for the future. Subcontractors' site personnel have responsibilities to their own company and to the main contractor, involving both lateral and downward communications.[1]

Spoken communication provides an important connecting link on all civil engineering projects. It encompasses face-to-face conversation, indirect telephone calls and dictated messages. Face-to-face communication can be very effective if carefully considered and clearly expressed. People are inclined to communicate more freely in the absence of a permanent written record, but the latter can also be a disadvantage. Management is able to determine the views of participants, but personnel may soon forget most of what is said.

Spoken communication can take the following forms:

(1) individual directives, such as engineer's instructions to the contractor;
(2) one-to-one discussions such as those that take place in an appraisal of a member of staff;
(3) manager to a group of people, as in a project briefing; and
(4) group discussions as occur in site meetings.

Communication Problems

Many organisational problems stem from failures of communication. Break-downs may occur through faulty transmission or receipt of messages or because of the messages being incorrectly interpreted. Fryer[1] has identified the following possible causes of communication failure.

(1) *Poor expression*: The communicator fails to formulate his message clearly, possibly because of difficulty in self-expression or limited vocabulary, or because of lack of understanding on the part of the recipient. People often fail to write or to speak clearly and simply. The contents of many reports and letters are obscured by long-winded and irrelevant words and phrases.
(2) *Excess of information*: It is common practice for management to produce and receive too much information. All communications should be concise and devoid of unnecessary information, as illustrated in later examples. There is a limit to the amount of information that a person can assimilate at any one time.
(3) *Unsatisfactory choice of method*: The communicator should consider care-fully the best form of communication in the particular circumstances. The spoken word may be the quickest and most effective in an emergency situa-tion, possibly followed by a confirmatory letter. Written communications are often preferred as there is then a file copy for future reference. In some instances the subject may be difficult to describe and a simple diagram could be much more effective.
(4) *Incorrect interpretation*: There is often a considerable divergence between the outlook, experience and attitude of the communicator and the recipient. Hence messages may be subject to misinterpretation because the recipient interprets the message in the light of his own experience, attitude and expectations, and what he regards as important. Where the communicator intentionally drafts the message in a certain way to safeguard his own posi-

tion or fails to express himself adequately, then problems of interpretation are likely to arise.

(5) *Remoteness*: Problems of distance make communication more difficult by eliminating the possibility of speaking face-to-face. This prevents the use of non-verbal signals like facial expressions which assist with communication.

(6) *Differences of status:* Differences in status between the communicants can make for difficulties in communication. Personnel in lower positions sometimes experience problems in expressing themselves freely when contacting a person of higher status, particularly when they are reporting on difficulties encountered or lack of progress.

(7) *Personal feelings*: How a person feels about a message or the sender may influence the way in which he interprets its contents and deals with it. In face-to-face communication this may be discernible often through the lack of response of the recipient. A manager should be able to grasp the situation and take positive steps to ensure that the recipient fully understands what he, the manager, is saying, accepts it and acts on it.

Alternative Approaches to Written Communication

Written communication such as technical reports can be produced in a variety of ways and the following example illustrates three approaches to the same task — a brief report on the construction of a national water sports centre and country park. The first approach is very long and cumbersome and this is followed by two alternative and much shorter presentations.

"The National Water Sports Centre and Country Park at Holme Pierrepont, Nottingham, represents a microcosm of combined leisure activities that are being developed in other places as separate items of provision, with all the disadvantages that this implies. It compresses into 100 hectares several totally dissimilar elements in close juxtaposition, but without the separate uses conflicting with one another. It might be said that we have been stretching credulity to its limits in attempting to locate massed viewing banks and a 2000 m rowing course only 50 m from a fisherman's sanctuary on the River Trent. One feels however that it will provide useful lessons in showing how far this type of synthesis is possible.

Recreation is a particularly appropriate word in that an area has been fundamentally changed and the landscape recreated, bearing little resemblance to what was there either before or after earlier gravel workings. Two disparate settings are interwoven to provide a spectator arena with the flavour of Henley and stylised nature formed around reedy backwaters, to form a wonderful blend of settings and activities which is bound to provide a great attraction for many years to come.

In assessing the measure of success that has been achieved in establishing a National Water Sports Centre in a Country Park at Holme Pierrepont, one should bear the following points in mind.

(1) This has been a rush job, with the whole of the constructional work having taken just over 2½ years from commencement of siteworks to completion of the main buildings. This is an exceedingly short period of time for complex works of such magnitude, costing £1.2m (1973 prices).

(2) During the period of construction, the results of three detailed research programmes have had to be considered and then incorporated into contracts, necessitating in one case the development of new techniques to test the particular problem.

(3) Three major sports bodies have been directly involved with the project and have had to formulate national standards to an extent far outside their previous experience.

(4) The centre is a prototype — with all the attendant problems of one — in that three different sports are combined on one stretch of water for competition and year-round training, yet with total regard for the essentially dissimilar needs of a country park and all that it entails. This must surely represent a masterpiece of planning and development.

(5) The overall capital costs are still below those originally estimated some three years previously for a fully fledged centre of national and international significance."

An attempt is now made to reduce this long, verbose and repetitive account into a much more concise report but still containing all the essential details. The purpose of the communication and the needs of the recipient(s) must always be given full consideration.

"The National Water Sports Centre and Country Park at Holme Pierrepont, Nottingham, combines a variety of leisure facilities, including a 2000 m rowing course to international standards, on a 100 hectare site, previously occupied by gravel workings. It contains substantial viewing banks which merge into reeded lakes and landscaped areas of the country park and the amenities of the River Trent.

In gauging the measure of success achieved, the following points deserve consideration:

(1) A very large and complex constructional programme costing £1.2m (1973 prices) was completed in 2½ years.

(2) The results of three substantial research programmes were incorporated into the contracts during construction.

(3) Three major sports bodies involved with the project developed new national standards.

(4) The one stretch of water accommodates three separate sports for competition and year-round training operating harmoniously with and complementing the adjoining country park.

(5) The overall capital costs were less than the initial estimate."

This represents a much more concise report without omitting any significant matters. It is much easier to read and assimilate, and the length has been reduced from 417 to 167 words.

Yet another alternative approach is to produce the information in a series of short sentences as illustrated in the following example. This procedure is to be avoided as it does not read well, being very disjointed.

"The National Water Sports Centre and Country Park at Holme Pierrepont, Nottingham, is located on a 100 hectare site. The site is on former gravel workings beside the River Trent.
Its main feature is a 2000 m rowing course.
It combines three separate water sports.
The results of three research programmes have been incorporated.
The construction period was 2½ years.
Final costs were below the original estimate."

The ideal approach lies between the two extremes as illustrated in the second account. This provides a satisfactory compromise between the long, verbose account in the first report and the undignified, short, sharp bursts of the last approach.

Choice of Written Style

As described by Scott,[2] a tradition has developed in civil engineering of using verbs in the passive tense. For example, 'Your report has been considered . . . . ' and 'Observations have been received from the employer.' In contrast, the active tense conveys a greater sense of commitment and urgency, as is evidenced by 'The employer has made observations.' Excessive use of the passive tense leads to a mediocre style.

In the past, engineers almost invariably wrote in the third person, such as 'It is recommended that . . . . ' and 'In the opinion of the writer it is unlikely that . . . . '. The modern trend is to adopt a more direct use of the first person pronoun, such as 'We recommend . . . . ' .

The four main variations in writing style are:

(1) length of sentences, paragraphs and sections, desirably kept to a reasonable length;
(2) choice of words, preferably kept simple and avoidance of professional technical jargon as far as possible;
(3) use of verbs, with the active tense generally preferred; and
(4) use of pronouns − first or third person.[2]

Whichever style is adopted, the civil engineer should aim to help the reader to follow his statements with understanding and interest.

Principles of Letter Writing

There are certain commonly established principles that are generally applied to the writing of letters. For example, Scott[2] has emphasised how the practice of referring to a particular paragraph in previous correspondence can be irritating to the reader and is generally to be discouraged. A typical example is 'With regard to the comments in the third paragraph of your letter of 27 May, the situation is . . . .'. The reader is obliged to refer to the earlier letter, involving waste of time and energy. It is much better to remind him of the text and then briefly to provide him with the precise reference to which he can return if he so wishes. For example, 'In response to your query about the capacity of the trunk water main (your letter of 27 May, third paragraph), the position is . . . .'.

The recipient of a letter is likely to receive a considerable amount of correspondence and the civil engineer should endeavour to produce concise, informative and yet interesting letters to attract the attention of readers. Some matters requiring particular attention in letter production are as follows: spacing; position on the sheet; error-free typescript; consistent margins; breakdown into readable sections, possibly combined with subheadings; avoidance of excessively long paragraphs; use of indentation to distinguish secondary from primary matters; and desirability of numbering sections and/or paragraphs.

In a lengthy letter, it is often good practice to start with an overview of the matters under consideration, and to finish the letter with a summary of the main recommendations. The opening words often have a considerable impact and the writer can capitalise on this fact. The use of the words 'This is to . . . .' have a high awareness factor and generally elicit the required response from the reader. The more conventional approach is to begin with 'Thank you for your letter of . . . .' or 'I thank you for your letter of . . . .'. These words make less impression on the recipient, but do, nevertheless, tend to be expected by some readers.

Where the writer uses the first person such as I, me and my, excessively, he may unwittingly give the recipient the impression that he is more concerned about himself than the reader. Conversely, if he continually makes use of the third person, such as 'It is recommended that . . . .', his writing tends to become stilted and uninspiring. The writer should only make extensive use of the first person when he wishes to assert his authority, as when a consultant writes to a contractor.[2]

The positioning of pronouns will often have a significant effect on their impact, with the first and last points in a statement being the most emphatic. A professional letter should desirably finish with an element of goodwill and may contain an expression of interest in a subsequent phase of the works, where appropriate.[2]

COMMUNICATIONS BETWEEN PARTIES TO THE CONTRACT

The Employer's Brief and its Development

Large public and private employers normally prepare a written brief before they engage the engineer for a project. Sometimes the briefs are poorly prepared and it is well worthwhile spending considerable time and exercising great care in an effort to make the brief as precise and comprehensive as possible. It forces the employer to thoroughly think through his ideas about the project and this prevents the engineer from wasting time in considering matters on which the employer has already made up his mind.[3]

Many engineers undertake the subsequent design work in isolation from the employer. On submission of the completed proposals for approval, it may then take the employer a considerable time to understand them and they may contain certain details which could have been eliminated at an early stage if the employer had known what was being contemplated.

Some engineers invite the employer's representatives to attend meetings of the design team. This practice certainly improves the employer's knowledge and understanding of the project and can quicken the decision-making process. However, some engineers argue that it can inhibit designers in the performance of their work, lead to premature decisions and can be wasteful of time.[3]

Appointment of Engineer

The engineer will need to clarify the terms of his appointment in writing to avoid any future problems. Most civil engineers use a standard appointment form which simplifies the procedure and reduces the risk of omissions. The following letter illustrates one possible approach to an appointment enquiry. Letters often incorporate a file reference, which include the initials of the writer and the typist.

North West Water Authority   27 August 1985
42 North Street
Wentworth
Northshire

Dear Sirs

<u>Amblehurst Impounding Reservoir</u>

We thank you for your letter of 21 August.

We are pleased to hear that you are considering appointing us as engineers for the above project. You are most welcome to visit us to discuss the work or, alternatively, you might find it more convenient to meet our representative at the proposed site. We can supply you with details of some of the projects undertaken by this office so that you can visit them and see the types of work for which we have been responsible.

A copy of our engineer's appointment form is enclosed for your information. The precise terms of our appointment can be agreed between us if you decide that you wish us to undertake the commission.

Yours faithfully
Jones, Smithson and Partners
Consulting Civil Engineers

Appointment of Other Consultants

The engineer may recommend the employer to appoint other specialists to provide skills and advice in connection with certain aspects of the project. For instance, the engineer may recommend the appointment of a quantity surveyor to measure and value the work where he does not have the necessary in-house expertise. The following letter provides one method that can be adopted by an engineer when approaching a quantity surveyor.

Peterson and Partners   16 October 1985
Chartered Quantity Surveyors
23 Market Street
Shipworth

Dear Sirs
             Britshire Docklands — Industrial Development Phase 1

Our clients, Docklands Development Corporation, have requested that we appoint you to carry out the quantity surveying duties for the roads and services on phase 1 of the industrial development at Britshire Docklands. The project consists of the provision of roads, footpaths, sewers, water mains and ducts for other services. We anticipate that the contract will start in March 1986 and the contract period will be approximately 8 months.

Will you please let us have, by return of post, your acceptance of the appointment and inform us of the basis upon which you wish to calculate your fees.

Yours faithfully
Johnson and Sandby
Consulting Civil Engineers

Consultation with Other Bodies

The engineer will often have to consult with other bodies such as local authorities, statutory undertakings and local consultative groups. Some of the bodies have statutory powers and the engineer needs to seek approval for certain aspects of the works, while other organisations are set up to safeguard local interests and a consultative process is likely to ensue. The following letters illustrate possible approaches to a planning authority and an amenity society.[4]

The Director of Planning   4 December 1985
Castle Chambers
Silchester
Northshire

Dear Sir

                    Amblehurst Impounding Reservoir

We refer to discussions held in your Department with Messrs Peters and Jackson on 28 November 1985.

In accordance with the agreement reached at the above meeting, we have revised the proposals and enclose three sets of drawings, numbers AIR 105/1A to 6A. These are in substitution for the original drawings, numbers AIR 105/1 to 6, and are submitted for consideration by your committee.

We thank you for your co-operation in this matter and look forward to receiving planning approval on 18 December 1985.

If there are any further queries, we would appreciate a telephone call in order that they can be resolved quickly.

Yours faithfully
Jones, Smithson and Partners
Consulting Civil Engineers

The Honorary Secretary   28 November 1985
Northshire Conservation Society
12 Cathedral Close
Silchester

Dear Sir

                    Amblehurst Impounding Reservoir

We refer to our telephone conversation of 25 November 1985 and confirm your agreement with our suggestion that Mr Jones should address a meeting of members of your society concerning the above project at your premises on 18 December 1985 at 7.00 pm.

We welcome your offer to arrange for the provision of light refreshments which will help to create a pleasant and relaxed atmosphere, in which a full and frank exchange of views on the environmental aspects of the project can take place.

Yours faithfully
Jones, Smithson and Partners
Consulting Civil Engineers

Correspondence between Engineer and Employer

Extensive correspondence is bound to take place covering a wide range of matters, starting from the initial appointment through to design, contractual, financial and constructional aspects. It is not possible within the limitations of this chapter to cover all the various topics that could arise on a civil engineering project, but the four selected examples will help to show how these specific matters could be dealt with. The reader will also appreciate that there is no single way of dealing with this type of correspondence and that it becomes very much a personal matter and each engineer has his own particular style.[4] The first letter deals with the employer's comments on the outline proposals for a project.

North West Water Authority   6 November 1985
42 North Street
Wentworth
Northshire

Dear Sirs
                    Amblehurst Impounding Reservoir

We thank you for your helpful comments on the outline proposals for the above project.

There has been no difficulty in incorporating most of your suggestions into our revised scheme. The only matter which has caused problems is the suggested relocation of the spillway channel.

We suggest that the best way to resolve this problem is for Mr Jones to visit you for a thorough discussion. We will telephone you shortly to arrange a suitable date and time for a meeting.

Yours faithfully
Jones, Smithson and Partners
Consulting Civil Engineers

The next letter covers the employer's proposals with regard to a sub-contract.

North West Water Authority　　　　　　　　　7 February 1986
42 North Street
Wentworth
Northshire

Dear Sirs

<u>Amblehurst Impounding Reservoir</u>

We refer to our discussion of 6 February 1986 concerning the proposed employment of Joseph Sparks for the sub-contract for electrical works on the above project and we confirm our advice to you.

(1)　You have instructed us to obtain a tender from one firm only, namely Joseph Sparks, for the sub-contract works.

(2)　We have advised you that, in our opinion, this firm is unsuitable for this class of work, and that tenders should be obtained from three other nominated firms.

(3)　We have further advised you that in the event of the firm being nominated to do this work, the consequences could include additional expense, delays in progress and completion, and lower quality of work than anticipated.

(4)　We shall, of course, carry out your instructions but we can take no responsibility should the outcome prove unsatisfactory, as the action taken is contrary to our advice. As soon as a sub-contract is signed we shall proceed to administer the contract provisions with diligence and impartiality.

Please consider the matter once again and let us have your final instructions as soon as possible.

Yours faithfully
Jones, Smithson and Partners
Consulting Civil Engineers

The third letter relates to a design change requested by the employer and the engineer takes the opportunity to include a timely warning.

North West Water Authority   21 February 1986
42 North Street
Wentworth
Northshire

Dear Sirs

Amblehurst Impounding Reservoir

We thank you for your letter of 18 February 1986 requesting us to alter the line
and width of the approach road. The necessary design work is underway.

Although the alterations are relatively minor, they will require a certain amount
of redrawing and rescheduling.

We know, from our earlier discussions, that you appreciate the problems caused
by quite small changes of design and the possible repercussions in terms of cost
and programme time. The problems are likely to intensify in the later stages of
design.

Yours faithfully
Jones, Smithson and Partners
Consulting Civil Engineers

Copy: Quantity Surveyor

The fourth letter covers the important matter of a qualified tender and the
recommended action.

North West Water Authority   26 July 1986
42 North Street
Wentworth
Northshire

Dear Sirs

Amblehurst Impounding Reservoir

Following today's meeting at which tenders for the above project were opened,
we consider that it would be helpful to you if we listed the main points that we
made at the meeting so that you may have the opportunity to consider them
more fully.

(1) All tenderers were informed that the tendering procedure would be in
    accordance with the Instructions to Tenderers incorporated in the contract
    documents and the Notes for Guidance on the Preparation, Submission and

Consideration of Tenders for Civil Engineering Contracts, issued by the Institution of Civil Engineers. Each tenderer understands the full implications and has the right to expect that, having expended a considerable amount of time and money on preparing tenders, these requirements will be strictly applied.

(2) A tenderer who amends the form of tender, inserts qualifications or submits a late tender is seeking to gain an unfair advantage. As you know, J.B. Peterson was unwilling to withdraw his qualifications. If all tenderers had been given the opportunity to qualify their tenders as they thought fit, any common yardstick for assessing tenders would disappear.

(3) All other considerations apart, the adoption of a universally recognised system of tendering has a beneficial effect on the whole construction industry by keeping prices at a realistic level.

We therefore advise that the irregular tender submitted by J.B. Peterson be rejected and that the normal checking procedure be applied to the next lowest tender. In the absence of any significant problems, and in conjunction with the quantity surveyor, we would expect to submit an acceptable tender total for your approval.

Yours faithfully
Jones, Smithson and Partners
Consulting Civil Engineers

Correspondence between the Engineer and Contractor or Sub-contractors

On a substantial civil engineering contract a considerable amount of correspondence will take place between the engineer and the main contractor and sub-contractors. They will encompass a wide range of contractual and technical matters and the following two letters illustrate commonly adopted forms of approach. Further documentation will be illustrated later in the chapter.

Quick Construction Ltd   30 July 1986
Hays Wharf
Edwinston

Dear Sirs
                    Amblehurst Impounding Reservoir

The employer, North West Water Authority, has instructed us to inform you that your tender of 20 July 1986 in the sum of £3 246 620 for the above project is acceptable and we are preparing the main contract documents for signature.

It is not the employer's intention that this letter, taken alone or in conjunction with your tender, should form a binding contract. However, the employer is prepared to instruct you to commence siteworks and place orders for the materials required in the first month of your contract programme.

If for any reason the contract does not proceed, the employer's commitment will be strictly limited to payment for the operations listed above. No other work included in your tender must be carried out without a further written order.

Yours faithfully
Jones, Smithson and Partners
Consulting Civil Engineers

Copy: Quantity Surveyor

The Clifton Pipe Company   1 May 1986
Oatsford
Georgeshire

Dear Sirs

<u>Amblehurst Impounding Reservoir</u>

Your tender for the supply of pipework for the above contract has been received.

However, the tender cannot be considered in its present form as it does not satisfy the requirements of the contract. If you still wish to tender, please complete the Standard Form of Tender, a further two copies of which are enclosed, and return it to us not later than 15 May 1986.

Please note that the submission of the tender on your own office form or the insertion of your own special conditions, other than in the appropriate position on the Standard Form, will result in disqualification.

Yours faithfully
Jones, Smithson and Partners
Consulting Civil Engineers

COMMUNICATIONS BETWEEN SITE PERSONNEL

A variety of communications will be required between various site personnel covering a wide range of matters. Some will originate from the engineer and will be mainly directed at the contractor and may be in the form of letters or instructions. Others will operate in the reverse direction, where the contractor is seeking information from or action by the engineer, while other communications may be internal ones between employers and their employees. A number of examples follow to illustrate the general format and the kind of topics covered.

Communications from the Engineer

The contractor is responsible for the setting out of civil engineering work and, despite some commonly held views to the contrary, checking by the engineer or his representative does not relieve the contractor of his responsibility under the contract. The following example will serve to reinforce this aspect, and is a reply to the contractor who has requested confirmation that his setting out is correct.[4]

Quick Construction Ltd   3 September 1986
Hays Wharf
Edwinston

Dear Sirs

<u>Amblehurst Impounding Reservoir</u>

We thank you for your letter of 29 August 1986 requesting confirmation that your setting out work is correct.

The setting out of the works on the site is your responsibility under clause 17 of the ICE Conditions of Contract. We consider that you have been supplied with all necessary data to carry out this task.

Any inspections made by us and the taking of any dimensions does not relieve you of your responsibilities in any way. We do not confirm that your setting out is correct and any lack of comment is not to be interpreted as an indication of our approval.

Yours faithfully
Jones, Smithson and Partners
Consulting Civil Engineers

From time to time during the course of the contract, the engineer's site staff may be dissatisfied with the standard of materials or workmanship and require removal of the defective items. The communication often takes the form of an engineer's instruction as illustrated in the following example, with copies sent to all relevant site personnel.[4]

Ambleside Impounding Reservoir

Engineer's Instruction Nr 11   16 October 1986
to Contractor

The following materials are not in accordance with the contract and must be removed from the site in accordance with clause 39 of the ICE Conditions of Contract.

(1) Delivery of 10 500 class B engineering bricks to south-east corner of site on 9 October 1986

(2) Delivery of 20 m^3 of coarse aggregate to concrete batching plant on 10 October 1986

(3) Delivery of 120 m of precast concrete kerbs to northern end of approach road on 13 October 1986

Copies: Quantity Surveyor
         Site Inspector

On occasions the engineer will require work which has been covered to be opened up for inspection, particularly where he has doubts about its quality and is suspicious because of the haste with which the work was covered up thus preventing its inspection by a member of the engineer's site staff at that time. The following example illustrates a suitable approach.[4]

Ambleside Impounding Reservoir

Engineer's Instruction Nr 14   29 October 1986
to Contractor

In accordance with clause 38(2) of the ICE Conditions of Contract, you are required to open up for inspection the base to the abutment at the south-west end of the dam wall. I intend to be present with the site inspector and the work must commence at 9.00 am on 31 October 1986. Failure to comply strictly with this instruction will result in your having to bear the cost of opening up and the subsequent reinstatement, regardless of the outcome of the inspection.

Copies: Quantity Surveyor
         Site Inspector

The Contractor is required to have a competent and authorised agent or representative constantly on the site to superintend the works. Problems could occur if the authorised person was withdrawn from the site without notice and the following letter illustrates the kind of action that the engineer might take, far reaching though it is.[4]

Quick Construction Ltd   6 November 1986
Hays Wharf
Edwinston

Dear Sirs

<u>Amblehurst Impounding Reservoir</u>

Clause 15(2) of the ICE Conditions of Contract requires you to have a competent and authorised agent or representative, approved in writing by the engineer, constantly on the works. You notified us on 14 August 1986 that Rodney Johnson MICE was your agent and we approved the appointment.

It has come to our attention that Rodney Johnson is no longer on the site. Hence you are in breach of contract until you notify us of the appointment and identity of his successor. Work must cease until a satisfactory appointment is made and no claim of any kind resulting directly or indirectly from the cessation of work will be considered.

Yours faithfully
Jones, Smithson and Partners
Consulting Civil Engineers

Communications from the Contractor to the Engineer

The majority of communications from the contractor to the engineer request information or identify discrepancies in the documents supplied by the engineer. The following two letters illustrate the form and content of typical communications.[5]

Jones, Smithson and Partners   19 November 1986
Consulting Civil Engineers
26 Victoria Street
Bilchester

Dear Sirs

<u>Amblehurst Impounding Reservoir</u>

We hereby apply for the supply of steel reinforcement details for the wing walls to the spillway. You will see from the approved revised programme that we intend to start the erection of the spillway walls in the week commencing 26 January 1987. In order to allow sufficient time for ordering, delivery, cutting and bending of the steel bars, we shall require this information by not later than 1 December 1986.

Yours faithfully
Quick Construction Ltd

Recorded Delivery

Jones, Smithson and Partners   23 January 1987
Consulting Civil Engineers
20 Victoria Street
Bilchester

Dear Sirs
Amblehurst Impounding Reservoir

We have not received a copy of your interim certificate nr 3 nor payment for the work executed during November 1986, in accordance with clause 60(2) of the ICE Conditions of Contract.

We shall be grateful if you will send us a copy of the certificate by return of post and arrange for payment to be made to us including the appropriate interest calculated in accordance with clause 60(6) of the Conditions of Contract on the overdue payment.

Yours faithfully
Quick Construction Ltd

Copy: North West Water Authority

Communications to Employees

Employers will communicate with employees concerning very satisfactory achievement such as examination successes, awards for excellent workmanship, good performance and high rates of progress. Unsatisfactory aspects will also call for letters to employees covering such matters as low production, poor workmanship and bad timekeeping, in addition to redundancy. The following letter shows a common form of approach.[6]

Mr Harold Goldsmith   16 January 1987
39 White Street
Marthambury

Dear Sir

Your timekeeping and attendance record shows that over the past three years
you have been absent 13 times and late on 19 occasions without a reasonable
explanation. You were given a previous warning on 19 December 1986.

This letter is a final warning that unless there is a significant and lasting improvement, we will regretfully have to terminate your employment with the company.

Yours faithfully
Quick Construction Ltd

   At least one verbal warning must precede the final written warning. In addition the employer must keep a record of all warnings.[6]

SITE DOCUMENTS AND RECORDS

Site Records

The keeping of continuous and comprehensive site records provides an effective
means of controlling and monitoring all activities on the site. They have a vital
role to play in the assessment and settlement of disputes. They can take a wide
variety of different forms and the following list embraces most of the more
common records kept by the engineer's site staff.

(1) all correspondence between the resident engineer and the agent, including
     engineer's instructions, variation orders and approval forms;
(2) all correspondence between the engineer for the contract and the resident
     engineer, the employer and third parties;
(3) the minutes or notes of formal meetings;
(4) daily, weekly and monthly reports submitted by the engineer's site staff;
(5) plant and labour returns, as submitted and corrected where necessary;
(6) work records such as dimension books, timesheets and delivery notes;
(7) daywork records, as submitted and corrected where necessary;
(8) interim statements, as submitted and including any corrections, with
     copies of all supporting particulars and interim certificates;
(9) level and survey books, containing checks on setting out and completed
     work;
(10) progress drawings and charts and revised drawings;

(11) site diaries;
(12) laboratory reports and other test data;
(13) weather records;
(14) progress photographs; and
(15) administrative records, such as leave and sickness returns, and accident reports.[7]

Correspondence

All letters, telexes, drawings and other documents should be recorded as they are received or despatched, and all incoming documents should be date stamped. Verbal instructions to the contractor should always be confirmed in writing and also telephone conversations where they convey instructions or important information. Copies of all correspondence, whether in the form of formal letters or handwritten notes, should be carefully retained, along with old diaries, note books, field books and similar data.[8]

Clarke[7] has described how correspondence files and site diaries form the central feature of any record system. The original of every incoming letter and a clear copy of every outgoing, with any enclosures, should be placed in a file containing a suitable reference title. One approach is to set up a file for each section of the bill of quantities and the specification, and for important subjects such as land ownership and boundaries, statutory undertakings, site safety, programme, setting out, sub-contractors and suppliers.

Where it is difficult to determine the most appropriate file heading, or when letters cover more than one subject, extra copies should be made and suitably filed. Additional but rather time consuming and costly measures that are sometimes taken include the keeping of a register of all correspondence and making an extra copy of each outgoing letter which is kept in a file or letter book in date order.

Reports

A report is primarily a summary of information and the principal method of conveying information on site matters to head office, the employer and other parties.

Daily reports by inspectors, supervising the constructional work on site, form an important part of site communications. They are often prepared in duplicate books, with one copy handed in to the site office and the other retained by the inspector. These reports contain details of the work carried out, weather conditions, the number of contractor's employees engaged on the work being supervised, number and types of plant in use and hours worked, and details of any delays and their causes. Starts and completions of activities will be noted. After processing, the reports should be filed and stored neatly and chronologically for ease of reference. Technical reports may be prepared on laboratory tests and special reports on specific problem areas.[8]

Labour and Plant Returns

The contractor's labour and plant returns constitute another commonly employed form of written record. The contractor is normally required to submit at prescribed intervals, such as monthly, the number and categories of labour and plant engaged on the site. The engineer's site staff carry out checks on the information provided.[8]

Drawings

Drawings provide a convenient and effective way of recording the progress of construction on the site. The type of information recorded includes the date and extent of construction (overall dimensions), the results of any tests and location of materials tested, and dates of approval of work. Record drawings may be purpose-drawn or the standard construction drawings suitably adapted to record construction progress. As the work proceeds, it is often necessary to revise drawings, normally on the original negatives and add dates and new references, to show alterations resulting from adoption of contractor's alternatives, different ground conditions and engineer's variations.

Laboratory Tests

Laboratory reports and other test results are normally entered on standard forms and filed on a subject basis. Common tests include concrete cube strengths, earthworks density, compaction and moisture content, and analysis of bituminous products. On occasions, the information is more effectively presented diagrammatically as in the form of graphs for matters such as standard sieve analyses. Statistical analysis of data can encompass the determination of such parameters as range, standard deviation and coefficient of variation. The laboratory may also undertake the recording of rainfall, temperatures, wind speeds and tides.[8]

Photographs

It is good practice to take photographs of the main features of the project from the same position at regular intervals, often monthly, to provide an excellent record of progress throughout the project. These photographs are often supplemented by photographs of particular features such as a rejected section of honeycombed concrete, irregular brickwork, bank slippage and extent of flooding resulting from exceptionally heavy rainfall. The photographs should be taken with a good quality camera preferably using coloured film. All photographs should have details of the date, subject, and position and direction from which it was taken recorded, usually on the back of the print. All negatives should be referenced to their prints, recorded and filed.

Diaries

Diaries are indispensable as they provide a complete narrative of the progress of the works and the activities of the engineer's site staff. The diary entries collectively supply comprehensive information on all aspects of the work and also permit cross-checking to elucidate disputed statements. A common type of diary is the self-carbonating duplicate variety with numbered pages. The absence of printed dates avoids any restriction on the length of entries. A line is drawn at the close of each day's entry and the next entry follows below, so that no further information can be inserted subsequently.

A diary provides a factual record of events on site, discussions with the agent's staff and other personnel, instructions issued and weather conditions. All entries must be accompanied by details of the time, location and personnel involved. Engineers and technicians must devote considerable amounts of time to the collection of appropriate information and its entry in their diaries. Rough notebooks and pocket tape recorders are useful for recording basic data and reminders. The resident engineer should check diaries regularly to ensure that they are up to standard.

Inspectors will need to record details of the deployment of plant and labour, movement of materials and progress of, and any problems associated with the work. Probably the best procedure is to use daily diary sheets subdivided into headed columns to receive information on time, location, activity, plant and labour, and general comments.

Assistant resident engineers or section engineers use the diaries to prepare a summary of the principal activities on both a weekly and monthly basis, which form a chronological review of the work in progress or completed.

The originals and carbon copies of diaries should not be stored together on the site because of the risk of loss. The carbons should be collected and bound together each week for despatch to head office. Completed diaries and inspectors' daily sheets should be stored in a locked, fire-resistant cabinet.[7] Figure 6.1 illustrates extracts from a site diary.

Variation Orders

Variations may be required to deal with variable site conditions, non-availability of materials or for other causes. On occasions an urgent decision has to be made by the engineer and he will probably initiate the variation order verbally on the site or over the telephone, to be subsequently confirmed in writing.

Under clause 51(2) of the ICE Conditions of Contract, provision is made for the contractor's agent to produce his own confirmation which, if not corrected

JONES, SMITHSON & PARTNERS		SITE DIARY
PROJECT: Ambleside Impounding Reservoir	NAME: P.T. Hinds	Nr. 1965

16.2.87

08·00 — Very cold with slight overnight frost.
Ground still saturated and unsuitable for earthworks.
Visited excavation at S.E. end of dam and met J. Martindale
(Sub-agent) and N. Tomlinson (Materials Engineer). Inspected the
exposed base and agreed that an area 6·5 m × 3·5 m in the S.W. corner
was very soft and should be removed and replaced with crushed
stone. A Hymac 590C tracked back-acter was available in the
vicinity and could be used for this purpose. Ordered J. Martindale
to proceed - formal variation order to follow. Arranged for an
inspector (T. Johnson) to attend and measure the depth of excavation
to a sound bottom.

08·55 — Returned to site office along approach road. No work in progress on
spillway. Pipes being laid between valve chambers 6 & 7.

14·40 — Prolonged showers. Considerable run-off into excavations at S.E. end
of dam. Toured area with T. Johnson. Excavations to dam abutment
now severely flooded and cut-off ditch overflowing. Record photograph
taken of flooded excavations. J. Martindale has arranged for 2 pumps
to be delivered to the site this afternoon. Warned him that the base
after it is dried out may not be suitable to receive the blinding course.
One articulated dump truck (Volvo DR 860) broken down.

15·50 — Piling rig being moved into position near the centre of the dam.

16·20 — Pipes laid between valve chambers 6 & 7 now submerged. T. Johnson
requested to keep the pipework under observation when the water level falls.

17·00 — Checked barriers and lights.
Prepared last week's progress report.

18·00 — Left site. Moderate rainfall.

P. T. Hinds

Figure 6.1 Extracts from a site diary

or repudiated by the resident engineer, becomes as effective as a variation order
issued by the engineer. These confirmations of verbal instructions (CVIs) require
careful scrutiny to ensure that they do not contain inaccuracies, differences of
emphasis or additional content. The following examples show how significant
differences can easily arise between the contractor's confirmation of verbal
instructions and the resident engineer's intended variation order.

CVI nr 14

The face of the concrete retaining wall adjoining the pumphouse is to be bush hammered to expose the aggregate.
Payment to be made on a daywork basis.

Variation Order nr 116

The following additional work is to be carried out.
Subject to a satisfactory trial panel, measuring 2 m x 2 m, in a location to be agreed, the face of the southern boundary wall adjoining the pumphouse is to be lightly bush hammered to expose the aggregate. The bush hammering is to extend from 150 mm above finished surface level to 50 mm below the coping.
Payment is to be based on the bill rate for item 28/D.[7]

Samples for Testing

It is advisable that a suitably completed form accompanies every sample that is sent to a laboratory for testing, and that a copy of the completed sample form and the test results are retained on the appropriate site file. Table 6.1 illustrates a typical samples for examination form.[9]

Delivery of Materials

The contractor will have a set of standard letters relating to such matters as delivery of supplies of materials which have not left the supplier, tracing materials delayed in transit, and advising the supplier of damaged goods and requesting replacement. The following example illustrates the type of letter that the contractor might send to the supplier in the latter case.

Roadmaking Supplies Ltd   9 January 1987
14 Roslyn Street
Charlesworth

Dear Sirs
                    Order No. 2039 — Precast Concrete Kerbs

We have to advise you that 18 of the kerbs received on 5 Janauary were badly damaged. We shall require these to be replaced urgently and shall be obliged if you will notify us of the date of despatch.

The carriers have been advised and we shall be pleased to receive your instructions with regard to the disposal of the damaged items.

Yours faithfully
Quick Construction Ltd

Table 6.1 Samples for examination form

BLOXFORD COUNTY COUNCIL

SAMPLES FOR EXAMINATION   SA/G2

To: The Scientific Adviser
    Scientific Branch
    Room 343, County Hall

Sample(s) submitted by: Department . . . .

                           Section   . . . .

                           Site   . . . .

Description of sample(s)   . . . .

                                      . . . .

                                      . . . .

                                      . . . .

Manufacturer . . . .

Brand . . . .

Contractor . . . .

Specification requirements   . . . .
   (if any)
                                      . . . .

Information required   . . . .

                                      . . . .

Name of officer requiring report   . . . . Telephone Nr . . . .

FOR OFFICE USE   Signature of officer
                            submitting sample . . . .

Lab. Nrs . . . .   Date . . . .

Employment

Many employers have a set of standard procedures relating to the advertising, selection and appointment of employees and a number of standard forms which help in maintaining a consistent approach and avoiding the omission of important facts.[6] Table 6.2 illustrates a typical employment application form for a trade operative and table 6.3 is a statement of particulars of engagement to ensure that a successful applicant is made fully aware of all the relevant factors relating to his appointment.

PURPOSE AND CONDUCT OF MEETINGS

Purpose of Meetings

During the course of a contract, a variety of meetings will take place in site offices, on specific parts of the works and in suppliers' premises. Some may be called at short notice to resolve a problem on the site, while others will be formally arranged at regular intervals and are generally concerned with co-ordination and progress. The main objective of all meetings is to come to a decision, although supplementary aspects such as the exchange of information, generation of ideas and discussion of problems may also be important. Meetings can, however, fail to achieve these objectives through over-formality, ineffective chairmanship, failure to concentrate on key issues or an antagonistic attitude by one of the parties.

Clarke[7] has described how informal meetings are generally concerned with the removal of unacceptable work or materials or the continuance or discontinuance of a particular method of operation. A note in the participant's diary may be sufficient to record the incident and the action taken. Where, however, there is a dispute over facts or liability, then the resident engineer must notify the agent of the time and place of the meeting, the names of those invited and the substance of their discussion. Meetings with sub-contractors in the absence of the main contractor have no contractual standing.

From time to time, it may be necessary to convene a formal meeting to discuss a specific matter which has become important to the progress of the project. Clarke[7] has classified them into three main categories.

(1) discussions between senior members of the site organisations, often involving the resident engineer and the agent;
(2) meetings with the main contractor and a sub-contractor; and
(3) meetings involving a third party, such as a statutory undertaker.

Franks[10] has suggested that the adoption of site meetings is a symptom of poor information flow from the design team to those who have to construct. He

Table 6.2 Form of application for employment

SURNAME (block letters)　　　　　　　　FIRST NAME(S) (block letters)

Present Address

Date of birth　　　　　Married or Single　　　　Registration number
　　　　　　　　　　　　　　　　　　　　　　　(if registered disabled)

TRADE (and/or special skill and　　　　　　　National Insurance Nr
　　details of training)

PREVIOUS EMPLOYMENT (over past three years)

Name of firm and site　　Type of work done　　Dates (from/to)　　Reasons for
　　　　　　　　　　　　　　　　　　　　　　　　　　　　　leaving

I certify that the foregoing particulars are correct and accept that my previous employers
may be asked for a reference.

Signed　　. . . . .　　Date . . . .

COMPANY USE ONLY

Particulars and date of offer:

Table 6.3 Statement of particulars of engagement

PARTICULARS OF TERMS OF EMPLOYMENT PURSUANT TO SECTION 1
OF THE EMPLOYMENT PROTECTION (CONSOLIDATION) ACT 1978

1. Name of employer   . . . .
2. Name of employee   . . . .
3. Job title   . . . .
4. Date of commencement of this employment . . . .
5. Your period of employment immediately preceding this employment does/does not qualify as 'continuous employment' as defined in Schedule 13 of the Employment Protection (Consolidation) Act 1978.
   Your period of continuous employment began on . . . .
6. During this employment your rates of remuneration, pay day, hours of work, holidays, public holidays and holiday pay, payment during absence from work due to sickness or injury, and other terms and conditions incorporated in your contract of employment, will be those appropriate to the work and the workplace where you are for the time being employed in accordance with the provisions of the following documents:

    (a) Your pay slip, pay envelope, or employer's wages register where your rate of wages if stated.
    (b) The Constitution and Working Rule Agreement of the Civil Engineering Construction Conciliation Board for Great Britain and any supplementary site procedure agreement.
    (c) The Annual Holidays with Pay Agreement in the Building and Civil Engineering Industries.
    (d) The bonus or incentive scheme, if any, applying to your work or workplace.
    (e) The employer's disciplinary and other written rules, if any, applying to your work or workplace.

7. There are no pension provisions in the terms and conditions of employment.
8. You may be required to transfer from one workplace to another in accordance with the provisions of the Working Rule Agreement.
9. You are entitled to receive, and are required to give, notice of termination of employment in accordance with the provisions of the Working Rule Agreement.
10. Any matter which may give rise to a grievance relating to your employment with the employer or relating to a disciplinary decision of the employer should, in the first instance, be taken up verbally with your immediate supervisor. Subsequent steps in the procedure are as prescribed in the appropriate document referred to in clause 6.
11. A copy of the Working Rule Agreement and of the other documents referred to in this statement, will be available for your inspection at the office of the workplace at which you are for the time being employed, or will be made available to you on request, and the employer undertakes to ensure that any future changes in the terms of these documents will be entered in such copies, or otherwise recorded for reference, within one month of the change.

I acknowledge receipt of a copy of the foregoing statement, together with a copy of the Company Safety Statement.

Signature of operative . . . . Date . . . .

Source: T.J. Gallagher Industrial Relations on Site.[6]

further argues that the use of site meetings as a means of reporting progress is also questionable, on the premise that the more time a person spends in reporting on what he is doing, the less time he has to carry out positive actions. Many engineers and contractors would dispute this line of argument.

Project Meetings

The most important meetings on a civil engineering project are the regular project meetings, sometimes termed site meetings or progress meetings. They are normally held at monthly intervals and provide the opportunity for a regular, comprehensive reappraisal of the project. These meetings are usually chaired by the engineer or a senior member of the head office staff and a person of similar status will represent the contractor. They permit a full and frank discussion of the contract, the giving of early notice of disputes which may not be capable of resolution at site level and the receipt of any legitimate complaints against the supervisory team.[7]

Fryer[1] has listed the main functions of project meetings as follows:

(1) to ensure that the contractor and other team members understand the project requirements and have the opportunity to check contractual, design and production details and to request clarification or information;

(2) to ensure that proper records are kept and contractual obligations met;

(3) to compare progress with targets and agree on any corrective action;

(4) to discuss problems, such as delays or sub-standard work which may affect the quality, cost or timing of the project;

(5) to ensure that sub-contractors agree on the action necessary to meet their obligations; and

(6) to check that variations are confirmed in writing and that work is recorded and agreed.

Conduct of Meetings

The chairman of the meeting, possibly in consultation with the secretary, ensures that dates for meetings are fixed, venues reserved and all participants notified. The chairman will approve the agenda and usually gives members the opportunity to suggest items for inclusion. He will also ensure that all relevant papers are sent to members well in advance of the meeting so that they all arrive adequately prepared. It is not unknown for an unscrupulous chairman to manipulate the agenda by contriving a protracted argument on a minor item in the early part of the meeting and subsequently to introduce an important matter at the end of 'any other business', as members are preparing to leave the meeting. A chairman should desirably be impartial, reasonable and responsive. Formal minutes or notes will be taken of the main points discussed and decisions made.

The chairman should plan the discussion around the agenda to enable members to make a positive contribution and to ensure the smooth and rapid progress of the meeting. On occasions, discussion on a controversial subject is far too long and the purpose of the discussion may become obscure. For example, contractor's senior staff could meet to consider whether or not to tender for a particular project and finish up discussing the method of execution of the contract.[2]

The main objective should be to reach unanimous assent in minimum time. An efficient chairman will at the outset of a meeting determine the purpose of the meeting, its direction, the limits of discussion and the timescale.[2]

Agendas

A formal agenda should be prepared for each project meeting to provide a sound basis for discussion at the meeting. It is usually formulated around a series of standard main headings, such as the major divisions in the bill of quantities, supplemented by relevant sub-headings, as illustrated in table 6.4.

Minutes

The minutes of a meeting normally record the following matters.

(1) the date and location of the meeting;
(2) those present;
(3) separate, numbered minutes of the matters discussed recording briefly:
    (a) subject
    (b) decision with action to be taken
    (c) name(s) of person(s) to undertake action;
(4) date, time and place of next meeting; and
(5) distribution list.[10]

A possible format and approach to the preparation of a formal minute are illustrated in the following example.

2.1 Review of progress for August

Mr. Johnson referred to the exceptionally heavy rainfall in August which had disrupted road and bridge construction and prevented the start of major earthmoving work. The RE said that the rainfall figures kept by the site laboratory would be compared with the Meteorological Office's local records for the past ten years. He referred to his comments at previous meetings concerning the unrealistic start date shown for earthworks in the contractor's programme. Mr. Miller asked whether the contractor was seeking an extension of time under clause 44 of the Conditions of Contract. Mr. Johnson replied that this was under consideration. The RE stated that details of any

Table 6.4   Extracts of an agenda of a project meeting

HIGHBURY BYPASS CONTRACT

Project meeting nr 6 to be held in the resident engineer's office at 2.00 pm on Wednesday, 3 September 1986

AGENDA

1. Apologies for absence

2. Minutes of the last meeting

3. Matters arising

4. General progress
   4.1   Review of progress for August
   4.2   Interim assessment of contractor's claim for extension of time.

5. Preliminaries
   5.1   Failure to keep public highways free of mud
   5.2   Inadequacy of storage huts

6. Site Clearance and Fencing
   6.1   Removal of trees subject to preservation orders
   6.2   Inadequate fencing between bridges 2 and 3

7. Drainage
   7.1   Delay in supply of land drains
   7.2   Delay in completion of surface water outfall to River Swale

8. Earthworks
   8.1   Contractor's proposals for method of working and for protecting the earthworks during the winter
   8.2   Contractor's request for information concerning variations to slipways (VO nr 61)

15. Nominated Sub-contractors
    15.1   Contractor's report

16. Accomodation Works
    16.1   Request for new brick boundary wall to Bogside House

17. Statutory Undertakers
    17.1 Diversion of overhead electricity lines between bridges 3 and 4
    17.2 Programme of Water Authority's works

18. Landowners and Insurance Claims
    18.1 New claims notified to contractor
    18.2 Encroachment on land to Jordson's Plantation

19. Safety
    19.1 Illumination of warning signs at Intersection 3
    19.2 Support to deep trench excavation

20. Any other business
    20.1 Proposed visit of Institution of Civl Engineeering Surveyors

21. Date of next meeting.

---

claim for an extension should be submitted promptly as the heaviest rain had been two weeks earlier and further delay in submission of details was not justified. The agent agreed to submit all available information within a week.[7]

Implementation of Decisions

At the meetings views are exchanged, proposals generated and decisions made. It still remains for the decisions to be implemented. The chairman of the meeting, possibly assisted by the secretary, will be responsible for ensuring implementation. The minutes will record who is to take the appropriate action and all participants will receive copies of the minutes. The action taken will be monitored at the next meeting under matters arising.[10]

REFERENCES

1. B. Fryer. *The Practice of Construction Management.* Collins (1985)
2. B. Scott. *Communication for Professional Engineers.* Telford (1984)
3. R.O. Powys. Improving communications in the building industry. *The Building Economist* (Australia) (June 1981)
4. D. Chappell. *Contractual Correspondence for Architects.* Architectural Press (1983)
5. V. Powell-Smith and J. Sims. *Contract Documentation for Contractors.* Collins (1985)

6. T.J. Gallagher. *Industrial Relations on Site.* Construction Press (1984)
7. R.H. Clarke. *Site Supervision.* Telford (1984)
8. J.W. Watts. *The Supervision of Construction.* Batsford (1980)
9. Greater London Council, Department of Civic Architecture and Design. *Handbook for Clerks of Works.* Architectural Press (1983)
10. J. Franks. *Building Sub-contract Management.* Construction Press (1984)

7 Measurement and Valuation of Work

This chapter covers the code of measurement and its application, taking measurements on site and keeping of essential records, evaluation of work on a dayworks basis, the making of interim and final valuations, valuation of variations and cost control by both the engineer and the contractor.

MEASUREMENT CODE AND ITS APPLICATION

Permanent Works

The principal code for the measurement of civil engineering work in the United Kingdom is the *Civil Engineering Standard Method of Measurement*[1] (CESMM). The majority of bills of quantities for civil engineering projects are prepared in accordance with this code, and they provide schedules of items giving identifying descriptions and estimated quantities of the work classified in the manner prescribed in the code. This procedure ensures a reasonable standard of uniformity in the approach to the measurement of these works.

The item descriptions for permanent works shall generally identify the component of the works and not the tasks to be carried out by the contractor. Clause 5.8 of the CESMM contains provisions which are more difficult to implement, in that the bill is to distinguish between those parts of the work of which the nature, location, access, limitation on sequence or timing or any other special characteristic is thought likely to give rise to different methods of construction or considerations of cost. Similarly, clause 5.10 requires all work to be itemised and described in accordance with the work classification contained in the CESMM, but adds that further itemisation and additional description may be provided if the nature, location, importance or any other special characteristic of the work is thought likely to give rise to special methods of construction or considerations of cost. In both cases, the views of the person preparing the bill of quantities, and the contractor who carries out the work could diverge significantly as to whether or not any special characteristics apply to the particular items of work. The principal objective is that the CESMM work classifications provide a minimum detail of description and itemisation on which contractors can rely, but encourages greater detail in non-standard circumstances, particularly where cost significant aspects are involved.[2]

Method-related Charges

The engineer enters specified requirements in the bill of quantities covering such items as accommodation and services for, equipment for use by and attendance upon the engineer's staff; testing of materials and works; and temporary works. The contractor, in his turn, is given the opportunity to enter method-related charges for items which are not directly related to the quantities of permanent work. These items include accommodation and buildings, services, plant, temporary works and supervision. The temporary works can encompass a wide range of diverse operations, such as traffic diversion and regulation, access roads, cofferdams, pumping, de-watering, access and support scaffolding, pits and hardstandings. The contractor must distinguish between time-related and fixed charges and fully identify the items inserted. With time-related items the contractor is to enter full descriptions of the items, including time and cost elements, to facilitate the subsequent costing of the work.

The items entered as method-related charges are not subject to admeasurement, although the contractor will be paid for these charges in interim valuations in the same way as he is paid for measured work. In the absence of variations ordered by the engineer, the sums entered against method-related charges will reappear in the final account, and will not be changed as a result of the quantity of work carried out being different from that originally estimated by the tenderer. The contractor is not obliged to construct the works using the methods or techniques listed in his method-related charges, but he will nevertheless be paid as though the techniques indicated had been adopted. For example, if the contractor inserted charges for a concrete batching plant and subsequently used ready-mixed concrete, the appropriate interim payments will be distributed over the quantity of concrete placed.[3]

Measured Work

The standardised approach introduced by CESMM does create some problems for the contractor in pricing billed items of measured work. On occasions associated work items are included in a bill term without the need for specific mention. For example, a typical earthwork bill item could read as follows:

E424   General excavation in natural material to a maximum depth not exceeding 1–2 m.

The rate would have to include the following items additional to cubic excavation, which are deemed to be included by rules in class E.

(1) Additional excavation needed for working space and removal of existing services.
(2) Upholding sides of excavation.

The contractor's estimator must also experience pricing difficulties when faced with a wide range of manhole types, sizes and connections where the enumerated approach to the measurement of manholes prescribed by CESMM is not sufficiently sensitive. Brick facework is not measured 'extra over' as in building work, and this makes pricing more difficult, since it involves deducting the common bricks which are displaced by facing bricks.

MEASUREMENT OF WORK AS EXECUTED

Where the engineer requires any part of the works to be measured he shall give reasonable advance notice in writing to the contractor, who shall either attend or send a qualified agent to assist in making the measurement. If the contractor fails to attend the measurement or send a representative, then the measurement made by the engineer or his representative shall be taken as being the correct measurement (clause 56(3) of the ICE Conditions). It is important that any delegation of the engineer's duties under this sub-clause and any limitations placed on them, shall be fully particularised and notified in writing to the contractor.

The main criteria to be observed in the measurement of civil engineering work are now described under the appropriate work sectional headings.

Earthworks

With excavation work, in addition to measuring the quantity of excavation to be carried out on the site, it is necessary to record the type of excavation (general excavation, foundations, cuttings or dredging) and the type of material being excavated (topsoil, rock, other natural material or artificial material). Items for foundation and general excavation are also classified according to the relevant CESMM maximum depth ranges below the commencing surface, which must be determined and recorded. Disposal of excavated material is a separate cubic item.

It must be borne in mind that if the contractor stockpiles without being instructed to do so, he will not be entitled to additional payment, even though it might have been difficult to avoid it, as in the case of excavated material to be subsequently used as fill. The quantities of material excavated or used as filling are measured nett, without any allowances for shrinkage, bulking or waste, with the exception of additional filling resulting from settlement of or penetration into underlying material in excess of 75 mm in depth — a difficult provision to apply on the site, requiring careful observation and measurement.[3]

In filling items, it is also necessary to distinguish between filling to structures, forming embankments, general filling and filling to a stated depth or thickness. The latter classification applies where the material is of uniform total compacted depth or thickness such as in drainage blankets, topsoiling, pitching and beaching. Filling to structures would include such work as filling around and over concrete storage tanks. Backfill to working space is not, however, measurable.

The trimming and preparation of excavated surfaces at an angle exceeding 10° to the horizontal are categorised into three groups (10° to 45° to horizontal, 45° to 90° to horizontal and vertical).

Suitably positioned ground levels should be taken over the site and compared with those shown on the project drawings, as these will form the starting point for the measurement of subsequent excavation work. With large areas of excavation, weighted grids or Simpson's rule may be used as the mode of measurement of the excavation. Some useful examples are given in *Civil Engineering Quantities.*[3]

Concrete Work

The volume of concrete provided on the project will be measured, with each mix kept separate. The placing of concrete requires much more extensive measurement, distinguishing between the different types of concrete (mass, reinforced and pre-stressed), different locations, and different structural members of varying size ranges, to take account of the different rates inserted in the bill. It should be borne in mind that the cost of placing concrete can be affected by the height above ground, position on plan, density of reinforcement, restrictions on access, limitations on pouring, exceptional curing requirements and related aspects, and these all require attention at the measuring stage, particularly where any variations to design have occurred during the course of the contract.

Columns and piers attached to a wall are measured as part of the wall, and beams attached to a slab as part of the slab except where they are expressly required to be cast separately. On the other hand, concrete in suspended walls and slabs less than 1 m wide or long, is measured as concrete in columns and beams respectively. When measuring quantities of concrete, no deductions are made for reinforcement, pre-stressing components, and most rebates, grooves, throats, fillets, chamfers, internal splays, pockets and holes and the like, and no additions are made for small nibs or external splays.

Formwork is measured to the surfaces of all *in situ* concrete which require temporary support during casting, and this excludes surfaces of concrete which are expressly required to be cast against an excavated surface. Formwork is measured to the upper surfaces of concrete inclined at an angle exceeding 15° to the horizontal. Formwork can be measured by length, for concrete components of constant cross-section. Typical examples are walls, columns and beams, where the formwork can normally be reused several times without major dismantling.

Where formwork is to be left in position for design purposes, it becomes part of the permanent works and is measured as such. The surface finish provided to the concrete will be recorded when measuring the formwork. With curved formwork, each different radius and shape of formwork will be measured separately.

Some joints to concrete works are measured by area; these include formwork, and the supplying and fitting of filler material. Items measured on a linear basis include water stops.

With *reinforcement* different materials and sizes are kept separate. Bars exceeding 12 m in length before bending need recording separately in multiples of 3 m.

Pre-cast concrete components as erected will be checked against the billed descriptions. The cost of the larger special components is influenced considerably by the shape, size and number required of each type.

Brickwork, Blockwork and Masonry

Walls of brickwork, blockwork or masonry are measured on their centre lines, normally by measuring on the outside face of the walls and adjusting for the corners. The quantities will be in m^2 if not exceeding 1 m thick, while thicker walls are given in m^3. No deductions are made for holes or openings with a cross-sectional area not exceeding 0.25 m^2. With walls of composite construction, the different materials are measured separately.

Vertical straight walls are distinguished from vertical curved walls, battered straight walls, battered curved walls, piers and columns, and these will each carry different rates in the bill of quantities. Surface features, such as copings, sills, corbels, band courses and plinths are measured separately as linear items. Fair faced work is measured in m^2 on the appropriate face of the walls.

Piling

Some piling activities are very appropriately covered by method-related charges. For example, the transport of piling plant and equipment to the site and the erection of stagings is normally covered by fixed charges and the operation of piling plant by time-related charges, when it will be necessary to record the time for which the plant is in use on the site.

Each group of items for cast-in-place concrete piles comprise: (1) number of piles, (2) total concreted length of piles and (3) total depth bored or driven. While those for reformed concrete and timber piles consist of: (1) number of piles of stated length and (2) total depth driven.

The section characteristics of the various piles must be checked and recorded. These embrace the mass/metre and cross-sectional dimensions for steel piles, and cross-sectional dimensions or nominal diameters for other piles.

With interlocking steel piles, both the driving and the materials are measured by area, which is determined by multiplying the mean undeveloped horizontal length of the pile wall by its depth. Interlocking steel corner, junction, closure and taper piles are measured as linear items.

Timberwork

The measurement of timberwork, such as in wharves, jetties and fendering, is relatively straightforward with the decking measured in m^2, subject to a void allowance of 0.5 m^2, and components by length in metres. The nominal gross cross-sectional dimensions or thicknesses (unplaned), species of timber, type of impregnation and any special surface finishes are all noted. Metal fittings and fastenings are recorded numerically.

Roads and Pavings

The various courses of road materials in sub-bases and surfacings are each measured in m^2, recording the material and its depth. No deductions are made for manhole covers and the like less than 1 m^2 in area. Kerbs, channels and edgings and their beds and backings, are measured as linear items, noting the materials and cross-sectional dimensions.

Pipework

Pipework is measured in metres giving the nominal bore and trench depth ranges as listed in CESMM. Pipework items are comprehensive ones in that they include the following items, as listed in Section I, in addition to the provision, laying and jointing of pipes.

(1) Jointing material.
(2) Lengths occupied by fittings and valves and those built into chamber walls.
(3) Excavation of trenches.
(4) Backfilling of trenches with excavated material.
(5) Temporary support to sides of trenches.
(6) Trimming trench bottoms.
(7) Disposal of surplus excavated material.

Work is separated locationally to take account of differing working conditions, paying particular attention to work in roads, through back gardens, restricted access and working conditions, need or otherwise for trench support and working around existing services.

All fittings and valves are recorded numerically with sufficient particulars for identification. The descriptions of pipe fittings, such as bends and junctions include the nominal bore, material, jointing and lining particulars.

Manholes, other chambers and gullies are recorded numerically and are often identified by the type or mark number listed in the drawings and/or specification. Manholes with backdrops are separately classified.

Pipe crossings of rivers, streams and canals will be recorded where the width exceeds 1 m, and crossings of hedges, walls and fences are separately recorded numerically.

Where the excavation and backfill of pipe trenches, manholes and other chambers involve rock, mass concrete, reinforced concrete or other artificial hard material, these will be measured in m^3 to cover the extra cost involved. The quantity is calculated by multiplying together the average length and depth of the material removed and the nominal width of trench excavation as stated in the contract. Where no width is given in the contract, the rules for measurement in CESMM will apply.

MEASUREMENT AND COST RECORDS

General Procedure

The periodic measurement of work executed on site is necessary to compare progress with programme and to assess the payments due to the contractor. The measurements can be taken from record drawings or physically on the site. The quantities of materials stored on the site will be assessed by direct measurement and observation.[4]

Watts[4] has described how in assessing the amount of work completed for purposes of interim measurements, it is sometimes convenient to agree average cross-sections or volumes when computing the volume of earthwork executed, concrete poured and other commonly used operations. Another method of simplifying the interim measurement calculations is to compute accurately the final volume and to allow a percentage of that figure for each valuation. Whatever method is used, care must be taken to ensure that the cumulative approximate evaluations do not exceed the value of the accurate final total.

Graphs are often prepared comparing the actual progress of the works against the programmed requirements for the major elements of the project both in quantitative and monetary terms. Other records that are normally maintained encompass labour (type, hours and rate), plant (type, hours and rate) and materials (type, quantity and rate). These latter records assist in assessing the cost of new or varied works at current or contract rates, and checking the value of any claims or estimates submitted by the contractor, including the verification of daywork items.[4]

An example of recorded site dimensions follows in table 7.1 relating to a sewer and giving all the essential particulars relating to the size and type of pipe, number and size of junctions, depth stages in accordance with CESMM, location, and number of hedge and fence crossings. This information is most conveniently recorded in the form of a schedule.

DAYWORK

Nature of Daywork

Daywork is the method of valuing work on the basis of time spent by operatives, materials used and plant employed, with an allowance to cover oncosts and

Table 7.1 Sewer schedule

Location	Type and size of pipe (mm)	Length of pipe (m)	Number and size of junctions (mm)	Length of trench (m) 1.5–2 m (total depth)	2–2.5 m (total depth)	2.5–3 m (total depth)	Number of fence and hedge crossings
MHs 1–2	225 concrete	88.800	–	–	13.800 (field)	6.500 (road) 10.000 (verge) 58.500 (field)	1 fence 1 hedge
2–3	225 concrete	70.600	–	–	70.600 (field)	–	2 hedges
3–4	150 gvc	41.400	–	35.400 (field) 6.000 (road)	–	–	1 hedge
4–5	150 gvc	58.500	4 nr 100/150	58.500 (road)	–	–	–
5–6	150 gvc	54.150	4 nr 100/150	54.150 (road)	–	–	–
6–7	150 gvc	64.370	–	13.000 (road) 51.370 (verge)	–	–	–

profit. Under sub-clause 52(3) of the ICE Conditions of Contract[5] the engineer may order in writing that any additional or substituted work shall be executed on a daywork basis. The contractor is required to submit to the engineer for his approval quotations for materials before ordering them. This procedure could result in delays to the works giving rise to a claim for extension of time under clause 44 of the ICE Conditions.

The use of daywork as a method of valuing constructional work should apply only where the normal process of measurement and valuation at billed rates or rates analogous thereto is not practicable. This method can offer advantages to both the resident engineer and the agent, as a contractor who is satisfied with the operative method of payment is likely to co-operate more readily in the execution of a complex variation or one involving unpredictable factors. Lengthy disputes over valuations can be reduced by the payment of dayworks to a relatively straightforward analysis of site records.[6]

On the other hand, quantity surveyors often contend that work executed on a daywork basis is more costly than similar work undertaken at billed rates, principally because there is no incentive for the contractor to execute the works expeditiously. However, by its very nature much of the work in this category is likely to take longer to perform than seemingly similar billed items. Many contractors argue that even although they insert seemingly high percentages in respect of dayworks, these rarely cover the actual costs incurred. They further state that they would prefer to have as little work as possible undertaken on this basis. Admittedly, the work normally executed on a daywork basis is relatively complicated and is likely to require increased supervision and cause considerable dislocation of the organisation and programming of the works, apart from the incidental additional costs involved.[7]

Daywork Schedules

The CESMM[1] prescribes three alternative procedures for the evaluation of dayworks.

(1) A daywork schedule which permits the entry in detail of separate rates for the respective classes of labour, materials and plant. The schedule will also contain a statement of the conditions of payment to the contractor. For instance, the rates inserted by the contractor are often required to cover such costs as overhead charges and profit; site supervision and staff; insurances; holidays with pay; use and maintenance of small hand tools and appliances; non-mechanical plant and equipment such as ladders, trestles, stages, bankers, scaffolding, temporary track, wagons, skips and similar items, unless these are used or set up exclusively for daywork; and in the case of rates for mechanically operated plant coming within the general classification of plant, items such as consumable stores, fuel and maintenance may be inserted. When travelling allowances or costs, lodging allowances and any other emoluments and allowances payable to operatives at the date of submission of the tender are to be included, this shall be made clear in the statement accompanying the schedule.[8] A typical daywork schedule illustrating this approach is shown in table 7.2.

(2) Provision for payment at the rates and prices contained in the current Schedules of Dayworks carried out incidental to contract work issued by the Federation of Civil Engineering Contractors,[9] usually with provision for adjustment by percentage additions or deductions for labour, materials and plant, from those contained in the Federation schedules.

(3) The insertion of provisional sums for work executed on a daywork basis, comprising separate items for labour, materials and plant, with provision for the insertion of percentage additions, to cover the contractor's overheads, charges and profit, as illustrated in table 7.3. The percentage addition for labour could be in the order of 130 per cent or more, whereas the additions for materials and plant are each likely to be around 12½ per cent. The author believes that this method offers the greatest advantages, since it

Table 7.2 Daywork schedule

Description	Unit	Rate
Labour		
Ganger	hour	
Concretor	hour	
Banksman	hour	
Timberman	hour	
Carpenter	hour	
Crane driver	hour	
Fitter	hour	
Steel bar fixer	hour	
Pump attendant (day)	hour	
Pump attendant (night)	hour	
Diver, including all gear, pumps and telephone	hour	
Boatman	hour	
Materials		
Sand	m³	
Coarse aggregate	m³	
Portland cement	tonne	
Rapid hardening cement	tonne	
6 mm mild steel bars	tonne	
12 mm mild steel bars	tonne	
16 mm mild steel bars	tonne	
25 mm mild steel bars	tonne	
Hardcore	m³	
Shuttering timber	m³	
12 mm plywood	m²	
Plant		
50 mm diameter submersible pump, including 30 m of hose	hour	
75 mm ditto	hour	
30 m of additional hose	hour	
Compressor, including tools (up to and including 20 m³/minute)	hour	
Crane (up to and including 5 tonnes lifting capacity)	hour	
Crane (exceeding 5 and not exceeding 10 tonnes lifting capacity)	hour	
Excavator (up to and including 1.40 m³ capacity)	hour	
Dumper (up to and including 1.50 m³ capacity)	hour	
Bar bending machine (hand operated, for bars up to 25 mm diameter)	hour	
Concrete mixer (200 litres wet capacity)	hour	

Table 7.3 Provisional sums for daywork

Bill reference	Item description	Unit	Quantity	Rate	Amount £
	Provisional sums				
	Daywork				
A411	Labour	sum			50 000.00
A412	Percentage adjustment to provisional sum for daywork labour	%			
A413	Materials	sum			24 000.00
A414	Percentage adjustment to provisional sum for daywork materials	%			
A415	Plant				24 000.00
A416	Percentage adjustment to provisional sum for daywork plant	%			

directly influences the tender total, and thus maintains an element of competition while, at the same time, providing a widely known and accepted basis of computation which is easily implemented. It is advisable to add a non-standard item for supplementary charges as prescribed in section 4 of the Federation Schedules,[9] where expenditure on this class of daywork is envisaged. These charges embrace such items as free transport provided by contractors for operatives to and from site, sub-contractors' accounts, plant hirers' accounts, internal transport on site, welfare facilities, additional insurance premiums and watching and lighting.[3]

The basic prices for elements making up the provisional sums are likely to be assessed in the following way.[6]

(i) Labour is normally charged at the basic rates of pay at the date of execution of the work, together with overtime rates, bonuses and plus rates for skilled trades, as detailed in the Working Rule Agreement of the Civil Engineering Construction Conciliation Board.

(ii) Materials are usually charged at the nett prices paid by the contractor.

(iii) Plant is conveniently charged at the rates listed in the FCEC Schedule of Dayworks.[9]

Procedure on Site

The procedural requirements are prescribed in sub-clause 52(3) of the ICE Conditions of Contract,[5] and need to be adhered to very closely by the agent and his staff to prevent any subsequent forfeiture of rights to payment. The main requirements can conveniently be listed as:

(1) All invoices, vouchers, delivery notes and receipts must be retained for subsequent inspection.

(2) Quotations for all materials must be submitted to and approved by the engineer prior to orders being placed. This is particularly helpful where the materials are very expensive or where there is more than one acceptable source.

(3) Full but unpriced particulars of all labour, materials and plant employed on daywork shall be submitted to the engineer's representative (resident engineer), daily and in duplicate. The labour particulars shall include the names and occupations of all operatives so employed and the times they spent on the work. The materials and plant entries shall contain adequate descriptions of the items and the quantities used. When the particulars are agreed, the engineer's representative shall sign the lists and return one copy to the contractor.

(4) At the end of each month the contractor is required to deliver to the engineer's representative a priced statement of the labour, materials and plant used. This is normally acceptable in a summarised form based on the detailed lists and statements submitted previously. It is important that the contractor submits the necessary particulars punctually in accordance with the prescribed timescale. Where the submission of these particulars is impracticable, the engineer is empowered to authorise payment for the work either as daywork or at such value as he considers fair and reasonable.

In practice the continual submission of daily records proves very difficult to implement. The staff in a busy agent's office are often under considerable pressure to keep up-to-date with routine work. Hence they often experience difficulty in extracting the necessary information, entering it on the contractor's standard form and despatching it in duplicate to reach the resident engineer the next morning. The agent should ensure that sufficient resources are made available for daywork records to be delivered to the resident engineer within 24 hours of the completion of the work to which they refer. Memories are likely to remain reasonably retentive within this timescale and discrepancies or inconsistencies can be readily resolved by a site inspection. Some further flexibility on timing is required as, for example, at weekends, bank holidays or for other acceptable reasons.[6]

As highlighted by Clarke,[6] when work is executed at daywork rates, the resident engineer is effectively hiring the contractor's labour and plant, usually at hourly or daily rates. The cost of any inefficiency, abortive time or lack of concerted effort is accordingly borne by the employer, in contrast to the more usual situation where work is undertaken at billed rates. The resident engineer must, therefore, maintain effective operational control of all daywork activities, and his site team must be prepared to take immediate action if they observe uneconomic working or wasteful use of resources.

Daywork Sheets

The following guidelines provide a useful approach to the preparation of daywork sheets.

(1) Each daywork sheet should relate to a specific engineer's instruction, preferably to a variation order or site works order, and the reference numbers of these should be stated on the daywork sheet.

(2) Daywork sheets should be submitted by the main contractor to the resident engineer for signature, the day after the work has been executed. Nominated sub-contractors should first submit their daywork sheets to the main contractor who will then include his own attendances and submit adjusted daywork sheets to the resident engineer.

(3) It is essential that daywork sheets should properly describe the work done. Since the daywork account may not be prepared for some weeks or even months, exact, clear and concise descriptions of the work are essential, possibly even supplemented by sketches or photographs of the work done.

(4) It is important to record the extra cost of purchasing small quantities of materials and any extra handling costs. When plant on the site is being used uneconomically instead of bringing more appropriate plant to the site, the approval of the resident engineer should be obtained as special rates may apply.

(5) Sub-contractors must prepare their daywork sheets in a similar manner to those of the main contractor, showing operatives' names and occupations and the like.

(6) Time must be allowed for setting out, covering up finished work, clearing away rubbish and making good, and these costs may be higher than normal.

(7) Daywork sheets should be carefully filed in sequence, together with the relevant instructions relating to them.

(8) Once a daywork sheet is signed by the resident engineer as a true record of the work done, this may not be the end of the matter and so all relevant drawings, sketches, instructions, delivery notes, invoices and receipts should be retained for future reference.

(9) Gerrity[10] describes how the obvious may be overlooked when preparing daywork sheets. In particular, cognisance should be taken of any scaffolding, shoring, special sections of formwork with only one possible use, extra long barrow runs, protection of adjacent finished work, access problems,

difficulties in working in confined space, sharpening tools for breaking out concrete, disorganisation of following trades, and other significant matters.

Daywork Accounts

The daywork account is built up from the daywork sheets which are normally submitted on a daily basis for civil engineering projects. The site engineer or quantity surveyor concerned will check the daywork sheets against the engineer's instructions, and he will list any of them that will be superseded by measurement and those covered by work in the bill of quantities. He will also check them for any inconsistencies and will record any details that appear excessive for discussion with the contractor's representative.

The resulting particulars, after verification and inclusion in a priced monthly statement, can be tabulated in a daywork account of the form illustrated in table 7.4, where the percentage additions inserted by the contractor in the daywork schedule, or daywork provisional sums in the bill of quantities, have been applied. The daywork will subsequently be incorporated into a variation account.

INTERIM CERTIFICATES AND PAYMENTS

Monthly Statements

Contracts formulated under the ICE Conditions of Contract[5] are measure and value contracts. The resident engineer has a duty to determine by admeasurement the value of the contractor's work and this task is carried out by subjective assessment based on his professional judgement and the more straightforward enumeration and measurement, or a combination of the two as illustrated later in this chapter.

Under sub-clause 60(1) of the ICE Conditions, the contractor is required to submit to the engineer a monthly statement which, unless prescribed otherwise in the contract documents, must contain the following details.

(1) the estimated value of the permanent works executed up to the end of the month covered by the statement;
(2) a list and the value of goods or materials delivered to the site but not yet incorporated in the permanent works;
(3) a list and the value of goods or materials, listed in the appendix to the form of tender, not yet delivered to the site, but of which property is vested in the employer pursuant to clause 54 of the ICE Conditions; and
(4) estimated amounts to which the contractor considers himself entitled, covering such items as temporary works and constructional plant and, with the operation of CESMM[1], this may include method-related charges.

Table 7.4 Daywork account

Cowbridge Power Station	*Rate*	£	£
Substructure and Siteworks Contract			
Engineer's Instruction nr 14			
(issued 19 June 1986)			
Work completed 3 July 1986			
Daywork sheets nrs 7–13			

Description
Demolition of Stores marked '**X**' on
drawing CPS/9/2B

Labour			
Labourer: 188 hours	3.00	564.00	
Craft operative: 40 hours	3.60	144.00	
Machine operator: 26 hours	3.60	93.60	
Ganger: 29 hours	3.80	110.20	
		911.80	
Percentage addition	130%	1185.34	
			2097.14
Materials			
Class B engineering bricks: 1800	190.00 per M	342.00	
Portland cement: 1 tonne	56.00	56.00	
Sand: 3½ tonnes	7.20	25.20	
		423.20	
Percentage addition	12½%	52.90	
			476.10
Plant			
JCB excavator: 26 hours	5.20	135.20	
2 tonne dumper: 26 hours	3.80	98.80	
Concrete mixer: 20 hours	2.00	40.00	
		274.00	
Percentage addition	12½%	34.25	
			308.25
Total to Summary			£2881.49

No statement is, however, to be submitted where the contractor considers that the total estimated value will fall below the sum inserted in the appendix to the form of tender as being the minimum amount of interim certificates under sub-clause 60(2). Amounts payable in respect of nominated sub-contractors are to be listed separately in monthly statements, as the engineer may require proof of payment by the contractor of the certified sums to the nominated sub-contractors.

In valuing the permanent works, the contractor will follow the sequence of the work sections in the bill of quantities. He will compute the quantities of the main items of work and multiply them by the appropriate billed rates to produce the interim values.

The claims for payment for goods or materials should be supported by copies of invoices, and any amounts for packing, delivery or replacements not chargeable to the employer should be deducted. The invoice amounts should tally with accepted quotations, and any additions will require an explanation from the contractor.

In the case of goods or materials stored off the site, the contractor must be able to show conclusively that their ownership has been transferred to the employer by means of a legal process known as vesting.

The extras claimed by the contractor should be adequately described and supported by appropriate calculations, which could be derived from quantities and rates but are more often in the form of lump sums.[6]

The contractor's monthly statements need to be supplemented by calculations, dimension sheets and any other relevant data, showing how the quantities and sums in the statement have been obtained. In the case of claims for extra payment, the supporting details shall be sufficient to enable the engineer to fully investigate the claims.

The monthly statement and supporting information are normally submitted to the engineer in triplicate. One copy is filed, another is used by the engineer's site staff for checking purposes and the third copy is returned to the agent with any required amendments inserted.

Monthly Payments

Following delivery by the contractor to the engineer or the engineer's representative of the monthly statement, the engineer is required under sub-clause 60(2) of the ICE Conditions[5] to certify and the employer to pay the contractor all within 28 days of the receipt of the contractor's monthly statement. To achieve this it will be necessary for the contractor and engineer to work closely together both in the taking of measurements and the agreement of rates.

Amounts certified by the engineer for items listed in sub-clauses 60(1)(a) and (d), comprising permanent works, temporary works, constructional plant and the like, will be subject to retention, while amounts certified against items listed in sub-clauses 60(1)(b) and (c), consisting of goods or materials, will not be subject to retention. Retention is calculated at 5 per cent of the amount due to the contractor until a reserve has accumulated to the employer up to the following limits.

(i) where the tender total does not exceed £50 000 − 5 per cent of the tender total but not exceeding £1500; or
(ii) where the tender total exceeds £50 000 − 3 per cent of the tender total.

Retention provides a fund on which the employer can draw to pay for any remedial work which the contractor refuses to carry out, and it also acts as an incentive for the contractor to complete any unfinished work or repairs during the period of maintenance. One-half of the retention money is payable to the contractor within 14 days of the issue of the certificate of completion for the whole of the works. The other half shall be paid to the contractor within 14 days after the expiration of the period of maintenance. If outstanding work remains to be executed by the contractor, the employer can withhold payment of such sum as, in the opinion of the engineer, represents the cost of the remaining work.

Failure by the engineer to certify or the employer to pay in accordance with sub-clauses 60(2), (3) and (5), renders the employer liable to pay interest to the contractor on any overdue payment calculated at the current bank rate plus ¾ per cent.

Action by the Engineer's Site Staff

The resident engineer and his staff must be satisfied as to the accuracy of the contractor's monthly statement before certifying payment to the contractor. Where, for instance, the contractor is unable to substantiate any of the amounts included in the statement, the engineer will reduce the value of the certificate proportionately. A considerable measure of co-operation is required between the resident engineer and the contractor's representative in compiling and agreeing the monthly valuation. The resident engineer may delegate the detailed checking, measurement and agreement of the valuation to the quantity surveyor for the contract.

The amount to be included in an interim valuation for preliminary and general items may be determined by any one of the following five methods.

(1) A percentage of the total value of the preliminaries is included equivalent to the percentage of the value of measured work completed by the contractor.
(2) A similar percentage is included proportional to the aggregate value of measured work and nominated work completed to date.
(3) A percentage is taken proportional to the number of weeks the contract has been in progress in relation to the contract period.
(4) The value of each item in the preliminaries and general items is assessed separately on the basis of the expenditure incurred. This expenditure is

likely to be greater during the early weeks of a contract when, for example, huts, plant and temporary roads need to be installed before the permanent works can proceed.

(5) The value of each item of plant and temporary works is assessed separately and the remainder of the preliminary and general items are valued using one of the percentage methods, and this procedure provides a convenient and reasonably efficient approach.[11]

Where the bill of quantities constitutes an accurate estimate of the quantities of the contract works, it will provide a good basis for computing the value of the measured work for interim valuations, otherwise remeasurement will be necessary. Regardless of the quality and accuracy of the bill of quantities, the first task is to prepare a record of the work executed at the time of the valuation, preferably carried out jointly with the contractor's representative. This is mainly achieved by an inspection of the site, when the extent of the work performed is recorded in notebooks or on drawings or diagrams. The data so recorded can then be checked against the daily record charts, bonus measurements, materials record sheets and labour returns. Following the site inspection, the value of completed sections of the work can be assessed either by pricing the approximate quantities of the completed work, or by determining the percentage of the work completed in each section of the bill,[11] as illustrated in table 7.5. Regard will also be paid to agreed records of covered up work by site staff.

It is a mistake to attempt to be too precise in calculating the amounts for interim payments and a sense of proportion must be maintained. The aim should be to assess accurately the value of the major items and to compute reasonable figures for the many smaller items. The engineer can carry out an approximate cross-check by comparing the sums expended with the forecast of completed work at the date of the valuation as shown on the contractor's progress chart.[11]

The value of unfixed goods and materials can be computed by preparing a priced list of the major items and adding an approximate sum for sundry items. The invoice values of unfixed materials on site is normally accepted by the engineer unless he is not satisfied with their quality or method of storage, or is of the opinion that materials have been brought on to the site prematurely or in excessive quantities, possibly for subsequent distribution to other sites. The engineer should also satisfy himself that the supplier has been paid for the materials.

Where the contractor or sub-contractors request payment on account for fabricated materials or machinery stored at their works, the engineer should give very careful consideration to such requests, obtain the employer's consent and ensure that the employer's ownership is assured.

Insurances are normally added to the valuation by the inclusion of the appropriate percentage. All daywork accounts require careful checking to ensure that there is no overlap with any measured work. Where the contract is subject to a price fluctuation clause, allowance will be needed to cover changes in the cost of labour and materials, usually calculated on the basis of the price adjustment formula. The adjustments are generally made on the basis of Baxter indices for which the base date was 1970 (100) and the indices covering the principal labour and material items are published monthly by HMSO.[12] For example, the indices for sand and gravel were 814.1 in March 1985 and 838.7 in June 1985. The index figures from which the price fluctuations are calculated are first published as estimates and confirmed figures are normally available three months later, when the original sum will be subject to recalculation. Furthermore, variations, extras and contractor's claims all need considering and including in the valuation where appropriate.

Interim Valuations

A typical example of an interim valuation for a road contract is illustrated in table 7.5 to show the general format and approach, starting with the adjustment of preliminaries and general items and followed by the various work sections, nominated sub-contractors, retention and materials, less the amount previously certified to give the amount due to the contractor under this valuation. The contract period is 20 months.

FINAL ACCOUNTS

After the works have been completed and the maintenance period expired, coupled with the satisfactory execution of all remedial and outstanding work by the contractor, the engineer for the contract issues the maintenance certificate under sub-clause 61(1) of the ICE Conditions of Contract.[5] The maintenance certificate states the date on which the contractor completed his obligations to construct, complete and maintain the works to the engineer's satisfaction.

Following the issue of the maintenance certificate, arrangements can proceed for the preparation and settlement of the final account. The procedure is set out in sub-clause 60(3) of the ICE Conditions of Contract,[5] and the importance of this stage of the contract and its administration is duly emphasised by making the engineer, as opposed to the resident engineer, responsible for this work. Where the interim certificates represent a fair evaluation of all the work contained in the contract, including billed items, variations and extras, the settlement of the final account should not be too difficult. In practice many factors combine to prevent this ideal arrangement taking place.

Within three months of the date of the maintenance certificate, the contractor is required to submit to the engineer a statement of final account and supporting documentation, showing in detail the value of work done in accordance with the contract, together with all further sums the contractor considers due to him under the contract up to the date of the maintenance certificate.

Table 7.5 Interim valuation

BROADSTONE BYPASS CONTRACT

Contractor: Road Construction Ltd

Valuation for Certificate nr 8: 12 June 1986

	£	£
Bill nr 1 Preliminaries and General Items		
Performance bond	24 000	
Insurances	122 000	
Offices for engineer's staff – establishment	6 000	
Ditto – maintenance, 8/20 × £12 000	4 800	
Attendance upon engineer's staff, 8/20 × £15 000	6 000	
Testing of materials	5 000	
Traffic regulation – establishment	7 000	
Ditto – operation and maintenance, 8/20 x £22 000	8 800	
Pumping plant – establishment	5 400	
Ditto – operation and maintenance	6 700	
Ditto – standing by	3 300	
Site accommodation – establishment	15 000	
Ditto – maintenance, 8/20 × £16 000	6 400	
Concrete mixing plant – establishment	16 000	
Ditto – operation and maintenance, 8/20 × £20 000	8 000	
Hardstandings	9 500	253 900
Bill nr 2 Site Clearance		
Bill total	51 000	
Removal of 15 nr tree stumps, 0.5-1.0 m diam., @ £20	300	51 300
Bill nr 3 Earthworks		
General excavation of material for disposal, maximum depth 0.5-1 m, 19 500 m³ @ £6	117 000	
Ditto—maximum depth 1-2m, 13 600 m³ @ £6.50	88 400	
General excavation of material for reuse, maximum depth 0.5-1 m, 15 600 m³ @ £4	62 400	
General excavation of topsoil for reuse, maximum depth not exceeding 0.25 m, 7 200 m³ @ £3	21 600	
Trimming of slopes, 6 600 m² @ £0.50	3 300	
Filling and compacting 150 mm of excavated topsoil to slopes, 6 600 m² @ £0.50	3 300	
Imported hardcore, 15 000 m³ @ £8	120 000	416 000
Bill nr 4 Carriageway		
Granular base, 150 mm deep, 50 000 m² @ £3	150 000	
Concrete carriageway slab, 225 mm deep, 45 000 m² @ £12	540 000	
Steel fabric reinforcement, 45 000 m² @ £2.50	112 500	
Expansion joints, 20 000 m @ £1.50	30 000	
Precast concrete kerb, 14 000 m @ £10	140 000	972 500

Bill nr 5 Footpaths
    Granular base, 75 mm deep, 1600 m² @ £1.50   2 400
    Bituminous macadam basecourse, 50 mm deep,
    1200 m² @ £4.00   4 800
    Bituminous macadam wearing course, 10 mm deep,
    1200 m² @ £1.50   1 800   9 000

Bill nr 6 Bridges
    Bridge 1
    Bill total £230 000 − 20 per cent complete   46 000

Bill nr 7 Culverts
    Bill total £670 000 − 25 per cent complete   167 500

Bill nr 8 Surface Water Drainage
    Bill total £240 000 − 40 per cent complete   96 000

Bill nr 9 Retaining Walls
    Bill total £94 000 − 20 per cent complete   18 800

Bill nr 10 Fencing
    Chain link fencing, 900 m @ £15   13 500
    Wood post and rail fencing, 480 m @ £8   3 840
    Chestnut pale fencing, 660 m @ £6   3 960   21 300

Bill nr 11 Dayworks   14 400

Variation nrs 1–13   12 670

Nominated sub-contractor:
Electrics Ltd − street lighting   78 000
Add for profit, 5 per cent   3 900
Attendance   6 200   88 100

       2 167 470
Less retention (3 per cent)   65 024

       2 102 446
Materials on site   77 554

       2 180 000
Less total of certificates 1–7   1 825 000

Total amount due   £355 000

Most of the information should already be in the engineer's possession in the form of the following documentation:

(1) the contractor's valuation of the measured work as submitted in monthly statements;
(2) the records and other cost data relating to variations which have usually been submitted to and discussed with the resident engineer; and
(3) the full and detailed particulars of all claims which should have been delivered as early as possible to the engineer's site staff, as prescribed by the Conditions of Contract.

In his statement of final account, the contractor will take the opportunity to substitute accurate amounts for earlier approximations, correct any discernible errors and, in the case of claims covering activities which continued up to the end of the construction period, it is probable that he can now provide all the necessary supporting financial data.

Since much of the work is often remeasured on completion, the services of a quantity surveyor can prove valuable. In the case of new or varied work, it is necessary to establish rates for items of work which did not appear in the original bills. The ICE Conditions of Contract (sub-clause 52(2)) prescribes that these rates shall be fixed by the engineer. In practice considerable discussion often takes place between the resident engineer and the contractor's representative, in order to establish fair and reasonable prices. The items for which rates are to be fixed normally fall into three categories.

(1) items to be priced *pro rata* to the contract rates;
(2) items to be priced at rates based on new build-ups which conform to the original method of rate fixing; and
(3) items to be priced at rates which are based on the original billed rates but adjusted to take account of the altered nature of the work or the changed conditions under which it is carried out.[11]

The valuation of variations will be considered in more detail in the next section of this chapter.

The work of nominated sub-contractors is normally remeasured on completion, preferably with a representative of the sub-contractor present. It is good practice to agree any new rates before the sub-contractor renders his account to the contractor. The contractor's profit will be computed at the same percentage as inserted in the contract bill, while the sum included for attendance in the original bill should only be adjusted when the quantity, quality or scope of the sub-contractor's work is changed. For example, a change in the cost of piling would not justify an alteration to the sum for attendance, unless the number of piles or their lengths are varied, or the method of piling is changed so as to vary the amount or form of attendance required.[11]

The final account normally consists of the following items.

(1) Preliminaries and general items as contained in the original bill with any necessary adjustments arising from changed circumstances or variations.
(2) Remeasurement of the works following the format contained in the original bill.
(3) Adjustment of prime cost items covering work carried out by nominated sub-contractors and goods or materials supplied by nominated suppliers.
(4) Daywork accounts.
(5) Labour and material price fluctuations where applicable.
(6) Variations based on engineer's instructions.
(7) Claims for loss or expense submitted by the contractor and agreed, possibly in a modified form, by the engineer.

It is often the settlement of varied work and claims which form the major tasks at this stage. Where agreement has previously been reached between the resident engineer and the contractor's representative, the values as agreed at an interim stage can be incorporated in the final account. It is almost inevitable that some rates and prices cannot be settled by agreement and must then be determined by the resident engineer. These rates will be considered by the engineer along with the contractor's final statement at the final account stage. The engineer will also consider contractor's claims and supporting particulars and the result of the resident engineer's investigations, before reaching a decision on them. The evaluation of contractors' claims is considered in chapter 8.

Within three months after receipt of the final account and all information reasonably required for its verification, the engineer is required to issue a final certificate stating the amount which, in his opinion, is finally due under the contract up to the date of the maintenance certificate. Ideally the final figure should be agreed with the contractor, who should issue a statement confirming this and certifying that there are no outstanding claims relating to the contract. The employer should be notified of any differences between the contract sum and the final amount.

VALUATION OF VARIATIONS

Variation Procedure

Under sub-clause 51(1) of the ICE Conditions of Contract,[5] the engineer has power to vary any part of the works, including temporary works, and ordered variations may include changes in the specified sequence, method or timing of construction. The engineer's powers do not, however, extend to variations of the terms and conditions of the contract.

Sub-clause 51(2) of the Conditions of Contract prescribes that the contractor should not make any variation unless he receives a written order or written confirmation of a verbal order from the engineer. The contractor may himself confirm in writing to the engineer a verbal order of the engineer and the latter must give written contradiction forthwith, otherwise it is deemed to be a written order by the engineer.

The engineer is not, however, required to issue a written order to cover variations in quantity of billed items (sub-clause 51(3) of ICE Conditions). He is, nevertheless, empowered, after consultation with the contractor, to vary rates or prices which have become unreasonable or inapplicable as a result of fluctuations in quantities (sub-clause 56(2) of ICE Conditions). Should the contractor disagree with the engineer's proposals, he should invoke the provisions of sub-clause 52(4)(a). This sub-clause recognises that in tendering, the contractor has no choice but to accept the billed quantity, but at the same time, the engineer cannot be precise in calculating the required quantities. In general an increased quantity should result in the more economic use of plant and a reduction in the billed rate. However, it may not always be so, particularly if more distant tips have to be used or plant requirements are changed.

A variation order should contain the following details:

(1) a concise but adequate description identifying the variation and including the relevant item number from the bill of quantities where applicable.
(2) The location in the works, including a cross-reference to any drawing on which it is shown.
(3) A description of the nature of the change, indicating whether it is an addition, omission or modification.
(4) Sufficient details to quantify the extent of the change.
(5) Details of timing where this is a significant factor.
(6) Any other information which the contractor may reasonably require to carry out the work.
(7) The basis of valuation, whether by billed rates or on a daywork basis.[6]

Valuation of Ordered Variations

The engineer is required to consult with the contractor prior to ascertaining the value of variations ordered under clause 51. The value of ordered variations is to be determined in accordance with the following principles as prescribed in sub-clause 52(1).

(i) Where work is of a similar character, and executed under similar conditions, to work priced in the bill of quantities, it shall be priced at applicable rates and prices.
(ii) Where work is not of a similar character or is not executed under similar conditions, the rates and prices in the bill of quantities shall be used as a basis for valuation so far as may be reasonable, otherwise a fair valuation is to be made.

Normally evaluation will be secured by agreement between the contractor and the engineer; failing this the engineer is empowered to determine the rate or price in accordance with the principles outlined and to notify the contractor.

Variations may render some billed rates or prices unreasonable or inapplicable and either the engineer or the contractor may give notice to the other that rates or prices should be varied, as prescribed in sub-clause 52(2). Such notice is required to be given before commencement of the varied work or as soon thereafter as practicable. The engineer shall fix such rates or prices as he considers reasonable and proper, presumably having regard to the component elements of the original rates or prices.

The majority of items of additional or modified work are covered by the procedure set out in sub-clause 52(1). The approach outlined in sub-clause 52(2) is normally only operative in two types of situation, namely:

(i) the valuation of omissions, and
(ii) the assessment of variations in which the payment of the full billed rate for additional work would provide the contractor with excessive profit.

The second situation arises when part of the billed rate relates to fixed overheads which do not increase *pro rata* with the quantity of work. For example, if the construction, maintenance and reinstatement of a temporary drainage channel is priced in the bill at a rate per week, doubling the length of time for which the channel is required does not double the cost to the contractor and the rate for the extended time requires adjusting.

On issuing a variation order, the resident engineer has a duty to evaluate its likely effect, as it could result in no cost difference, a cost saving or extra cost. It could be that a modification results in no delay, disruption or additional work, and should therefore have no financial consequences. For example, a minor change in the line of a road notified well in advance can fall into this category, as could the substitution of one type of road gully for another of the same cost, requiring no extra labour and ordered prior to the contractor making any purchases from the supplier.

When the contractor is ordered to omit work, there is sometimes a presumption that the contract sum can be reduced by deducting the value of the quantity deleted at the billed rates. It could, however, adversely affect the contractor as he may already have placed orders for materials which are subject to cancellation charges or the goods may have been delivered and payments made for them. Plant and other resources may have been assigned to the site, involving the contractor in direct expenditure. Preparatory or temporary works may have been executed in readiness for the cancelled work and other operations may have been programmed around it, which could have been carried out more economically had the agent received earlier notice. In these circumstances the contractor is entitled to payment to reimburse him for the direct loss and expense incurred.[6]

Most variation orders cover modified, new or additional work and involve extra cost which can normally be assessed in accordance with the procedure prescribed in sub-clause 52(1). Where the variation results in an increase in the quantity of an item in the bill, as for example the felling of six extra trees, it is likely that the billed rate can be used to value the work as it is of similar character and executed under similar conditions. However, when the work is not directly comparable to that priced in the bill, for example if the six extra trees were of a different size and type and located on another part of the site with some access problems, then the billed rate may be used as the basis for valuation as far as may be reasonable. This process requires consultation between the engineer's site staff and the contractor involving, for instance, the comparison of the billed work with the varied work and probably the analysis of the billed rate to enable a new rate to be determined. It should be possible to reach agreement provided the principle of analogous rates is accepted and the mutually agreed returns covering the varied work are available.[6]

It may be that the nature and scope of the ordered work is such that no appropriate billed rate exists. In such cases the Conditions of Contract require the resident engineer and the agent to consult in order to agree a fair valuation, which generally takes the form of the reasonable cost of the work, plus a percentage for profit. Where the contractor's agent considers the billed rate to be inappropriate, the resident engineer must give him the opportunity to put forward his proposals for assessing a new price, supported by all necessary particulars, including the build up of his contract rates. The resident engineer will check their accuracy and assess their relevance to the valuation.[6]

Where the parties are unable to agree on the method of valuation, the resident engineer must resolve it himself. If the contractor has failed to supply full particulars in substantiation of his price, the resident engineer is not obliged to present his own detailed calculations to the agent. The Conditions of Contract require the price or rate to be determined after consultation and the contractor notified. There is no requirement to show how the value has been calculated which could, if provided, encourage the contractor to start a protracted negotiation process. The resident engineer must resist the temptation to arrive at a valuation which unduly favours the employer, as he must be fair and impartial in all his dealings with the contractor.

Clarke[6] has demonstrated how executing work out of sequence, on a different scale to that depicted in the contract documents or in a different manner from that originally envisaged is likely to produce costs in excess of those attributed to the operation itself, because of the effect on other adjacent or concurrent activities. This is illustrated in the following example.

A contractor erecting a pumping station was instructed to change the base level of the pumping chamber after the excavation had been completed and trimmed. The only suitable machine available had to travel from the far end of the large site across an area which had become badly rutted by construction traffic in adverse weather conditions. Although the excavation and re-trimming work took only a few hours, the machine and operator were not available for any other work for a full day.

The valuing of variations may not be straightforward, and the resident engineer should give full consideration to all possible side effects, indirect losses, delay and disruption, and consequential costs. No alterations should be made to site records to make provision for these factors. If, for instance, excavation plant could operate at only 40 per cent efficiency then this should be stated and the resident engineer's confirmation requested. If operations elsewhere are delayed or deprived of supplies, the agent's estimate of the resultant loss should be submitted for the engineer's inspection and assessment. The submission and assessment of contractors' claims are examined in chapter 8.

FINANCIAL CONTROL OF CONTRACTS

Cost Control by the Engineer

The importance of budgetary control systems has increased considerably in recent years owing partly to the funding arrangements for public sector projects, partly to the reputation construction contracts have gained almost worldwide for over-spending, and partly to the employer's demands to start projects quickly, and consequently without adequate preparation.

The resident engineer, assisted by other site staff, often including a quantity surveyor on large projects, should at an early stage agree ground levels with the contractor and suitable arrangements for dealing with daywork sheets and claims for increased costs. An accurate record of all drawings, both original and revised, should be maintained and all variation orders costed and filed. Continuous records should be kept of all significant matters such as labour strength, plant in use, weather conditions and causes of delay, as these could subsequently have a bearing on the subjects of claims. Throughout the contract period, the site quantity surveyor or other responsible member of staff should maintain effective cost control arrangements by keeping a constant check on costs and by supplying cost advice to the resident engineer, in ample time for any necessary corrective action to be taken without adverse effects on the project.[13]

It is important to ensure that any variations, claims or extras do not raise the likely final account figure above the budget. It is particularly necessary to monitor the financial effect of variations and these should, ideally, be costed before they are issued. The resident engineer or his appropriate representative liaises with the contractor at regular intervals for the valuation of variations, to agree remeasured work and to discuss any claims submitted by the contractor and/or sub-contractors.

The contingency sum should be used only to cover the cost of extra work that could not reasonably have been foreseen at the design stage as, for example, extra work below ground resulting from poor ground conditions. Early consideration should be given to expenditure against provisional and prime cost sums, and the examination of sub-contractors' and suppliers' quotations.

The resident engineer, or the quantity surveyor acting under his direction, will normally produce monthly forecasts of final expenditure, and predict and monitor cash flow. The progress is best shown in the form of a graph with possibly a solid line depicting the anticipated rate of expenditure calculated from the contractor's master programme, usually of the characteristic S curve shape, with perhaps a broken line showing the actual rate of progress drawn by connecting the gross values of work done at each valuation. In this way the contractor's progress on site can be checked and the probable consequences of any delays to completion determined. These form an important aspect of the financial management of capital schemes. The receipt of periodic financial statements or reports by the employer enables him to anticipate his future financial commitments and to revise his capital budget where appropriate. They normally list authorised expenditure, details of savings and extras, actual expenditure, progress and other related matters.[7]

Should any of the cost information obtained prove unsatisfactory, such as the possibility of final expenditure exceeding the tender total, urgent action must be taken to rectify the situation.

Summing up, the employer should be informed of his financial commitments and when he will be required to make payments. The resident engineer and his staff must effectively control expenditure on variations, provisional and prime cost sums, set against quotations, quality control, completion to time and claims. The problem of major additional requirements not provided for in the initial contract and needing extra funding is one of paramount importance and affects all parties to the contract.[7]

The employer can reasonably expect to be supplied with the following information:

(1) An estimate of the final account at regular intervals during the contract period, preferably monthly.

(2) A comparison of the estimated costs with the total allocated financial resources.

(3) If the employer's total allocated financial resources comprise components from different resource bases, an allocation of the estimate between these bases.

(4) If the comparison between the final account estimate and the resource allocation is unfavourable, he will require an explanation and an indication of remedial action.

(5) An indication of when he will be expected to pay money and approximately in what amounts.

Cost Control by the Contractor

Cash Flow

The assessment of the profitability of a particular contract consists basically of knowing precisely the value of work executed at a specific date, compared with the actual costs incurred in achieving that value of work. The difference between the two figures will be the amount available to allocate to the off-site overheads of the company, to fund its working capital and make a profit. In an adverse situation the difference may show that off-site overheads are not being covered and that no profit is being made. In the worst situation, the actual costs of construction on site may exceed the value of the work that those costs have generated.[14]

Somerville[15] describes how the usual reason given for a company's difficulties is cash flow, whereas this is more often a symptom of the problem and not the cause. Many construction companies become insolvent through bad estimating and planning, ineffective contract control or inadequate site cost control. Cash flow may be defined as the actual movement of money in and out of a business. Within a construction organisation positive cash flow is derived mainly from monies received through monthly payment certificates. Negative cash flow is related to monies expended on a contract to pay wages, purchase materials and plant and meet sub-contractors' accounts and overheads expended during the progress of construction. On a construction project, the nett cash flow will require funding by the contractor when there is a cash deficit; where cash is in surplus the contract is self-financing. With contracts operating under United Kingdom standard conditions with retention funds and low percentage profits on turnovers, construction firms are frequently in financial deficit for much of the contract period.[16]

Cash flow problems can be reduced if effective procedures can be operated by the contractor in respect of the following matters:

(1) realistic monthly assessment of preliminaries from fully documented and priced preliminary schedules;
(2) increased costs under contracts with fluctuations kept up-to-date in monthly valuations;
(3) variations to the contract accurately assessed and included in valuations;
(4) daywork sheets completed and cleared for monthly payment;
(5) discounts and retention monies properly claimed against the contractor's own nominated sub-contractors and suppliers;
(6) collection of all monies properly due to the contractor; and
(7) ensuring that all claims for loss and expense are fully documented, properly presented and submitted as quickly as possible.

Site Cost Control

It is cost control in the context of profit or loss that is the primary concern of the contractor's quantity surveyor. In this capacity he works closely with the agent, who is monitoring performance, comparing it against pre-determined targets and taking remedial action where necessary. The factors to be controlled include the tangible physical resources of operatives, materials, machines and sub-contractors. Equally important are the non-tangible items such as progress and productivity (time), cost (money), quality, safety, information, methods and the performance of subordinate management staff.[17]

The main sources of data available to the contractor's quantity surveyor are the contract bill of quantities, estimates of cost, method statement and the master programme. During construction these four data sources will be supplemented by interim valuations; up-to-date accounts of labour, plant, materials and sub-contracted work; salaries and all other site costs; and finally the programme of actual work executed compared with the assumptions upon which the tender was based. Supported by this back-up data, achievement of effective cost value comparison will involve the following activities.

(1) The calculation of the true value of work carried out on a cumulative basis to a prescribed cut-off date.
(2) The restatement of the true value of the work in terms that can be directly compared with costs. Some contractors refer to this restatement as the preparation of the earned allowances, whereby in relation to the work carried out, not more than a certain sum should have been spent on labour, plant and other components.
(3) Costs are collated to the same cut-off date as that adopted for the statement of true value, sometimes termed 'true selling value'. Costs will be adjusted to take account of liabilities for costs that have not yet been recorded but against which value of work has been taken, and for costs generated against which no value of work has yet been created.[14]

Supplied with these three sets of data, it is then possible to ascertain whether profit or loss is being made on expenditure against labour, materials, plant, sublet work and site overheads, and whether the level of contribution is better or worse than that upon which the tender was based. The comparison may be made in terms of the whole project or, if data in terms of both value and cost can be accurately sub-divided, into the construction elements making up the total project.

The computerised breakdown of interim payment applications into earned allowances is probably the most common application of computers to the financial management of contracts by contractors. A number of standard packages are available and, should the engineer's staff receive a payment application in the form of a computer print-out, they can be reasonably certain that earned allowances have been used in preparing the application.

As a management tool, the contractor's quantity surveyor is looking for consistent and inconsistent trends as between the value and cost of each of the earned allowances month by month. If a consistent trend is observed, for instance by regular over-spending on labour at a consistent level, then management will endeavour to identify some underlying reason. Examples include the all-in labour rate being higher than anticipated at the time of tender, or output being consistently lower than the constants upon which the tender was based. The inconsistent relationship between earned allowance and actual cost is more likely to have been caused by an isolated event, and again management will want to identify the reason and obtain a solution in each case. The key factor is to recognise that there is a problem at a time sufficiently contemporary with the event to be able to take effective action. On the cost side of the cost value comparison, it is essential that there is close liaison between the contractor's quantity surveyor and the site accountant on a very large project.[14]

Financial Reporting

Good management practice dictates that reliable and regular financial reporting is necessary to control a project effectively and reports should be produced ideally on a monthly basis. A basic financial report of a contract should contain:

(1) initial tender figures and expected profit;
(2) forecast figures at completion for value and profit;
(3) current payment application by the contractor;
(4) current certified value;
(5) adjustments to the certified valuation;
(6) costs to date and the accounting period in question; and
(7) cash received to date, retention deducted and certified sums unpaid.[18]

Cost Value Reconciliation

The cost and value of variations must be continually monitored and assessed. A contractor must ensure that he has clearly defined procedures for identifying variations and that he conforms fully with the requirements of the contract regarding notices and the supply of supporting information and particulars. After interim valuations have been made, the contractor should list all unagreed claims, variations, daywork, remeasurement, interest on overdue sums and any other disputed items. The contractor should also assess the progress achieved by monitoring the value of work carried out against that programmed. This can best be done by constructing an S curve of the forecast valuations against time and then plotting actual value against actual time.

Accurate recording of the cost of materials, plant, labour, site staff and overheads, and sub-contractors' work and claims is essential in cost value reconciliation. For instance, there is frequently a delivery charge for plant and this may not have been indicated on the initial order. This can include time for the plant travelling from its depot as well as the delivery charge itself. When operators are provided the question of overtime arises and also greasing or servicing time.

A monthly reconcilation of materials delivered to the site should be made against the quantities certified in the measurement. Allowance must be made for materials rejected, used on site but not measured as in strengthening temporary roads, used off site as in minor work for adjoining landowners, and materials that are stockpiled. If the quantity unaccounted for exceeds the estimated wastage allowance, further investigation is needed. Possible causes are errors in measurement, additional work being performed without supporting orders, or loss through unforeseen circumstances, such as excessive penetration of granular material into a very soft sub-grade.[6]

With regard to labour costs, site staff usually have the responsibility to complete timesheets indicating the number of hours worked by each operative and the amount of bonus earned in that particular period. Management will monitor the allocation of staff on the various contracts, and will record holidays, sickness or other reasons for absence and overtime payments, and check that staff and overheads charges are kept within the intended budget.

Residual credits may occur when materials or items of plant have been purchased for a certain contract but on completion still retain some foreseeable value which can be used on other future contracts. They do, however, require careful examination and evaluation.

REFERENCES

1. Institution of Civil Engineers and Federation of Civil Engineering Contractors. *Civil Engineering Standard Method of Measurement* (1985)
2. M. Barnes. *Measurement in Contract Control.* Institution of Civil Engineers (1977)
3. I.H. Seeley. *Civil Engineering Quantities.* Macmillan (1977)
4. J.W. Watts. *The Supervision of Construction.* Batsford (1980)
5. Institution of Civil Engineers, Association of Consulting Engineers and Federation of Civil Engineering Contractors. *Conditions of Contract for use in connection with Works of Civil Engineering Construction.* Fifth Edition (June 1973, revised January 1979)
6. R.H. Clarke. *Site Supervision.* Telford (1984)
7. I.H. Seeley. *Quantity Surveying Practice.* Macmillan (1984)
8. C.J. Haswell and D.S. de Silva. *Civil Engineering Contracts.* Butterworth (1982)
9. Federation of Civil Engineering Contractors. *Schedules of Dayworks carried out incidental to Contract Work* (1985)
10. H.B. Gerrity. Variation orders, site works orders and daywork records. *The Practice of Site Management.* Chartered Institute of Building (1980)

11. R.J. Marks, R.J.E. Marks and R.E. Jackson. *Aspects of Civil Engineering Contract Procedure.* Pergamon (1985)
12. *Price Adjustment Formula for Construction Contracts — Monthly Bulletin of Indices.* HMSO (published monthly)
13. I.H. Seeley. *Building Economics.* Macmillan (1983)
14. B. Meopham. Cost control for the civil engineering contractor. *Chartered Quantity Surveyor* (June 1983)
15. D.H. Sommerville. *Cash Flow and Financial Management Control.* Surveying Information Service. Chartered Institute of Building (1981)
16. B. Cooke and W.B. Jepson. *Cost and Financial Control for Construction Firms.* Macmillan (1979)
17. J.G. Gunning. Site management — mainly a process of control. *Building Technology and Management* (December 1983)
18. F.R. Barrett. *Cost Value Reconciliation.* Chartered Institute of Building (1981)

8   Settlement of Contractors' Claims

This chapter is concerned with the different types of contractors' claims, procedural aspects, the relevant contract conditions and their application, origination, preparation and assessment of claims, additional cost for loss or expense, disruption of work resulting from variations and the operation of liquidated damages.

GENERAL BACKGROUND TO CLAIMS

Nature of Claims

The construction industry covers a complex field of activity involving many operative skills and conditions which vary considerably from one project to another. Site and climatic conditions, market conditions, project characteristics and availability of resources are some of the variables, each of which can have a significant effect on the operation of the contract.

Most construction contracts make provision for these complexities and uncertainties by the inclusion of clauses permitting the contractor to claim for loss or expense resulting from specific contingencies. The ICE Conditions of Contract[1] attempt to clarify the contractual requirements and remove any ambiguities as far as possible. In the absence of these provisions, contractors would have to include in their tenders for many more uncertainties than they do now, which would result in a significant increase in tender totals. However, under the standard form of contract, the employer will only have to meet the cost of such contingencies if they arise and have been duly verified.

The term 'claim' as used in this context is a request by the contractor for recompense for some loss or expense that he has suffered, or an attempt to avoid the requirement to pay liquidated and ascertained damages, as described later in the chapter. It is in this light that claims should be viewed by both sides of the industry. Frivolous claims by contractors to redress the effects of inefficiency or profit shortfall are unlikely to receive sympathetic consideration by the engineer or employer. It has been justifiably argued that the term 'claim' should be used only in respect of fundamental breaches of the contract and that the remainder are contractual entitlements. A claim to be successful must be well prepared, based on the appropriate contract clauses and founded on facts that are clearly recorded, presented and provable.[2]

Action by the Contractor

The well organised contractor will be able to recognise the occurrence of events that are likely to result in his ultimately suffering loss or expense. A contemporaneous record of such events should be sent to the engineer as a basis of the claim. The late submission of claims by a contractor will inevitably result in difficulties, since neither the employer nor the engineer will have had the opportunity to check the factors giving rise to the additional expense, at, or about the time of their occurrence. A contractor can so easily fail to obtain reimbursement of monies to which he is entitled because of the late submission of claim or notification of his intention to submit one.[2]

The potential loss and expense must be clearly identified, quantified and valued. In addition, other parties to the contract must be convinced that they are valid claims and that the integral parts are claimable and correctly valued.

Robinson[3] has described how, depending on the quality of the records, a good, indifferent or bad claim is produced. Most poorly produced claims are the result of a hasty, last minute analysis of sketchy and incomplete records. The best prepared claims come from the management of contractors who appreciate that loss and expense situations are likely to arise on contracts, and accordingly establish procedures to identify and record all relevant background information and data, in order that an accurate and well-founded evaluation can be made promptly. The contractor should ideally have an efficient organisation which is continually looking ahead to identify and diagnose possible future problems.

The contractor must apply in writing for the issue of instructions, details, drawings, levels, or for the nomination of sub-contractors as appropriate. He must also give written notice to the engineer of any cause of delay in the progress of the work and written notification in respect of any claims that he is contemplating making in respect of variations or loss and/or expense. He must take positive steps to ensure that the engineer's instructions are issued in writing or verbal instructions confirmed. Alternatively, the contractor may confirm in writing to the engineer any oral order by the engineer and where the confirmation is not contradicted in writing by the engineer forthwith, it shall be deemed to be an order in writing by the engineer (sub-clause 51(2)). The contractor should ensure that the various certificates required under the contract are issued by the engineer, particularly in respect of completion, maintenance and extensions of time, to prevent any unnecessary problems arising in the future. It is always a better policy to avoid disputes rather than being involved in their settlement.

Action by the Engineer

Burke[4] has described how some architects, and it applies equally to engineers, appear to interpret any letter from a contractor asking for details or information as a prelude to a claim; it is, however, a contractual duty of the contractor to apply in writing for such details if they are not supplied to him. By so doing, he may remove the necessity for making a claim, if he receives the information promptly.

The engineer should recognise that most contractors will be looking for opportunities of submitting claims, whether fully justified or otherwise, especially in periods of economic depression, and should not take offence when adequate notice is given. It is in the interests of all parties to the contract to deal with extensions of time and evaluation of monetary entitlement at the time when the relevant events occur and the pertinent facts are still fresh in everyone's minds. Furthermore, their satisfactory resolution has the effect of keeping the contractor on target both for money and for time. Another advantage is that a contractor tends to keep his best management and employees upon those contracts that he knows have effective completion dates and cost targets and where claims will be dealt with expeditiously.[5]

Hughes[6] advises that whatever the merits or demerits, the engineer and/or his consultant quantity surveyor would be wise to encourage the contractor to keep him informed of anything that is happening or has happened involving the possibility of additional expense. In some cases the work in hand may be concerned or affected and the engineer may be able to take remedial action, where the fault rests with him or the employer, or avoiding or mitigating action, where the cause is one for which the employer has accepted the risk.

It is the responsibility of the contractor to formulate his claim in detail and to furnish the evidence on which the claim is based. A properly supported claim will be carefully examined to establish whether the facts are properly founded, whether the matters submitted equate to the circumstances provided for in the contract clauses and whether the amount claimed can be justified.[7] Trickey[8] has summarised the main actions of the engineer and/or quantity surveyor on receipt of a claim from the contractor, as determining:

(1) which clauses of the contract apply and how are they to be interpreted?
(2) what elements of cost are involved? and
(3) what is the monetary entitlement?

Many engineers dislike dealing with claims because:

(1) it occupies time and energy that could more profitably be devoted to other work;
(2) it implies increased costs which will displease the employer;
(3) it may imply some lack of care on the part of the engineer, which could be embarrassing and result in action by the employer against the engineer;
(4) they sometimes experience difficulty in handling claims in an effective manner; and
(5) should the engineer's decision be disputed, the matter in dispute may be referred to arbitration or courts of law.[9]

General Approach to Claims

In most cases when assessing a claim the engineer and/or quantity surveyor will require the master programme, which is often in the form of a bar chart, a method statement showing in general terms how the contractor intends to carry out his work, and a detailed breakdown of the costs of preliminaries and general items.[5]

Simonds[10] has set the scene in emphasising that contracting is a high risk business. Therefore the contractor at the time of tendering has the right to be able to plan and expect to proceed with his work in an orderly manner. The employer, on the other hand, has the right to expect experience and competence from the contractor. The contractor should not attempt to make the employer pay for his mistakes and the employer must not expect the contractor to bear the cost of errors or changes made by him or the design team. A contractor, when pricing a tender, is asked to take risks. When the contract is let and under-way, and the employer or his agents prevent the contractor from carrying out his contractual obligations properly and effectively, then he will probably find it necessary to make a claim.

TYPES OF CLAIMS

Broad Categories of Claims

There are three main categories of claim:

(1) *Contractual claims*

These are claims that are founded on specific clauses within the terms of the contract. This type of claim will be considered in more detail later in this chapter, with particular reference to the ICE Conditions of Contract.[1]

(2) *Ex-contractual claims*

These claims are not based on clauses within the terms of a contract, although the basis of the claim may be circumstances that have arisen out of the project and have resulted in loss or expense to the contractor. On occasions a sympathetic employer has settled an ex-contractual claim (a term disliked by lawyers), because the contract was on time although the contractor suffered exceptional misfortune.

However, in general these claims are unlikely to succeed as there is no contractual obligation for payment and any payments made are in the nature of *ex gratia* payments (act of grace). Typical examples are where late deliveries of materials by a supplier on a firm price contract resulted in substantial price increases on the materials, or where difficulty was experienced by the contractor in recruiting adequate labour and he was obliged to pay high additional costs to attract them.

(3) *Common Law claims*

Many of the clauses in the standard forms of contract are stated to be 'without prejudice to any other rights and remedies.' Such rights are established by taking action through the courts for damages for breach of contract, tort, repudiation, implied terms and other related matters.[2]

PROCEDURAL ASPECTS

On civil engineering contracts claims are submitted by the contractor for consideration by the engineer. The engineer has a duty under the contract to resolve disputes and consider claims in an independent and impartial manner without showing bias towards either party. The contractor must accept the engineer's decision on most matters until the completion of the works, subject to the arbitration provisions of clause 66. Whether the contractor's claim encompasses time or money or both, the engineer has to make a decision on its validity and quantification.[11]

Under the ICE Conditions of Contract a contractual claim may be submitted to the engineer at any time during the currency of the contract and up to three months after the date of issue of the maintenance certificate. The majority of claims on civil engineering contracts relate to adverse physical conditions and artificial obstructions (clause 12), extension of time (clause 44), delay claims arising out of clauses 7, 12, 13, 14, 27, 31, 40, 42 and 59B, ordered variations (clause 51) and valuation of ordered variations and rate fixing (clause 52).[11]

If a contractor intends to claim any additional payments he is required under sub-clause 52(4)(b) of the ICE Conditions to give notice in writing of his intention to the engineer as soon as reasonably possible after the happening of the event(s) giving rise to the claim, and should thereafter keep such contemporary records as may be reasonably necessary to support any claim he may subsequently wish to make. Without necessarily admitting the employer's liability, the engineer may upon receipt of the notice instruct the contractor to keep all necessary contemporary records. The contractor is required to supply the engineer with copies of the records as and when required (sub-clause 52(4)(c)).

The contractor shall as soon as is reasonable send to the engineer a first interim account giving full and detailed particulars of the amount claimed to that date and the grounds upon which the claim is based. This submission shall be updated at the intervals required by the engineer (sub-clause 52(4)(d)). If the contractor fails to comply with any of the provisions of sub-clause 52(4), the contractor is entitled to payment only to the extent that the engineer has not been prevented from investigating the claim.

The contractor is entitled to have included in any interim payment certified by the engineer, such amount in respect of any claim as the engineer may consider due based upon the particulars supplied by the contractor. If the particulars are insufficient to substantiate the whole of the claim, the contractor is entitled to payment in respect of such part of the claim as the particulars may substantiate to the satisfaction of the engineer (sub-clause 52(4)(f)).

APPLICATION OF CONTRACT CONDITIONS

Adverse Physical Conditions and Artificial Obstructions

Most civil engineering contracts involve a substantial proportion of work below ground. Under clause 12 of the ICE Conditions, a claim may be submitted as a result of adverse physical conditions or artificial obstructions encountered during the execution of the works, and which could not reasonably have been foreseen by an experienced contractor at the time of tender. This clause has wide ranging implications and the final decision on the validity of any claim rests with the engineer who, on occasions, in the author's experience, seems to adopt a rather harsh and uncompromising stance. There is provision for reference of disputes to arbitration but in practice the contractor rarely uses this option for a variety of reasons, including the high cost and the possible adverse effect on future invitations to tender.

A claim submitted under the provisions of clause 12 may be for additional payment to cover the cost of dealing with the conditions or obstruction(s). Sub-clause 12(3) prescribes that this shall include 'a reasonable percentage addition in respect of profit, and the reasonable cost incurred by the contractor by reason of any unavoidable delay or disruption of working suffered as a consequence of encountering the said conditions or obstructions or such part thereof'.

The type of adverse physical conditions that can occur may range from running sand, unpredictable water tables and subsidence to geological faults, while artificial obstructions could encompass old tips and foundations to unrecorded mine workings and underground services. From time to time attempts have been made to define 'running sand'. Some engineers consider it to be water-bound sand or sand that will not stand up unsupported. It is, however, generally accepted that the term applies to sand whose condition changes after being worked.[2]

Extension of Time for Completion

Under clause 44 of the ICE Conditions, the contractor can request and the engineer can grant extensions of time for completion of the works. It should be noted that an extension granted under this clause does not automatically entitle the contractor to any additional costs incurred by him during the period of extension.

Under sub-clause 44(1) the contractor is required to submit full and detailed particulars of any claim to extension of time within 28 days after the cause of the delay has arisen or as soon as reasonable thereafter. The engineer is required to grant such interim extension of time as he considers due (44.2) and to further assess any extension of time to which he considers the contractor is entitled at the due date for completion (44.3). A further review is required upon the issue of the completion certificate (44.4).

The engineer is bound to notify the contractor of any extensions that are granted and of any claims for extension of time to which he considers the contractor is not entitled. The engineer should also keep the employer fully informed of the decisions made and they may affect the payment of liquidated damages, as described later in the chapter.

Claims for Delay

Where it can be demonstrated by the contractor that the delays that have occurred cannot be classified as contractual risks and constitute genuine delays for which payment is specifically provided or implied in other contract clauses, then the contractor is entitled to payment in respect of the extra costs incurred. This category of delay arises when the contractor is unable to deploy his labour and plant in such a way that the intended output can be achieved as related to the critical path on the approved programme. The principal sub-clauses in the ICE Conditions which specifically refer to payment for delays are:

7(3)   Delay in issue of drawings or instructions.
12(3)   Adverse physical conditions or artificial obstructions.
13(3)   Engineer's instructions or directions to clarify ambiguities or discrepancies in accordance with clause 5.
14(6)   Delay in approval of proposed methods of construction or unanticipated limitations on design criteria.
27(6)   Delay resulting from variations covering emergency or other works under the Public Utilities Street Works Act.
31(2)   Delay caused by the provision of facilities for other contractors employed by the employer.
40(1)   Suspension of work.
42(1)   Late possession of site.
59B(4) and (6) Delay and extra cost on forfeiture of a nominated sub-contract.

Claims under Other Contract Clauses

There are a variety of other clauses and sub-clauses in the ICE Conditions, under which the contractor may be entitled to claim for additional payment. These are now listed to assist the reader in locating them.

18   Boreholes and exploratory excavation which constitute a variation.
22(2)   Employer's indemnity on damage to persons and property.
26(1) and (2) Giving of notices, payment of fees and conforming with statutes.
30(3)   Employer's indemnity on bridges and highways.

36(3)	Cost of tests not provided for in the contract.
38(2)	Cost of uncovering work and making openings.
49(3)	Cost of execution of work repair during period of maintenance.
50	Searches required by the engineer.
55(2)	Correction of certain errors in the bill of quantities.
56(2)	Increase in billed rate arising from change in quantity.
58	Adjustment of provisional and prime cost sums.
59(A)(3)(b)	Direction by the engineer in respect of nominated sub-contractors.
60(3)	Final account claims.
60(6)	Interest on overdue payments.

Variations

In respect of the pricing of variations, the engineer under sub-clauses 52(1) and 52(2) is given authority to determine certain prices. If the contractor intends to claim a higher price than that determined, he must under sub-clause 52(4)(a) notify the engineer within 28 days.

The time limitation is all important. After giving notice that he intends to claim under this sub-clause, the contractor must send to the engineer an interim account giving full and detailed particulars of the amount claimed to date and the grounds upon which the claim is based. The contractor shall update this information when the full facts and extent of costs are known.[2]

ORIGINATION OF CLAIMS

On-site during Course of Works

It is important that the occurrence that results in a situation where the contractor suffers loss or expense is identified soon after the event. The verification of the supporting information and facts, which are so important to the success of the claim, is very difficult in retrospect.

Members of the contractor's management team are likely to experience a variety of situations where certain aspects are not entirely satisfactory. The following examples will serve to illustrate some of these problems:

(1) the site surveyor observes an operation proceeding on site but cannot determine clearly how it is going to be reimbursed;
(2) the contracts manager experiences difficulty in obtaining information from the engineer despite a number of approaches; and
(3) the managing director is unhappy about the profitability of the project.

In each of these circumstances immediate investigation is warranted, to identify the cause and to determine whether it provides a realistic basis for a claim and, if so, to give notice of the claim.

In a contractor's organisation information regarding the actual cost of a contract compared with its estimated cost or value is often produced far too late. As a result any investigation into the cause of the loss is in retrospect, and hence the information is more difficult to establish, collate and record. A detailed cost study should be carried out on the more important elements of the work. One approach is to relate actual cost to budgeted cost from the estimator's records on carefully selected elements where this form of comparison can be done readily, albeit approximately, and regularly on site. Alternatively, a properly integrated bonus scheme, to achieve the optimum level of productivity, will provide much of the information needed.

An accurate register of drawings is vital with their suffixes, date of receipt and the main changes introduced. Site minutes and notes of all meetings must be carefully filed after being scrutinised by the contractor to check their accuracy. Any doubtful aspects should be queried in writing and raised at the next site meeting. Site agents and foremen should keep comprehensive diaries as these may subsequently prove invaluable in supporting claims and even more so at an arbitration.

Where it is evident that the contractor will sustain a loss that can form the subject of a claim, full records of all pertinent facts should be kept and notice given to the engineer. Here again an effective costing system is vital.[2]

Chappell[9] has shown how contractors' claims may be generated by a variety of factors, including:

(1) discrepancies between drawings, specification, schedules and the bill of quantities;
(2) late or hurried preparation of detailed information;
(3) multiplicity of engineer's instructions;
(4) constant revisions to drawings;
(5) changes of mind on the part of the employer; and
(6) poor co-ordination between two or more parties involved in the project.

Periodic Checks by the Management Team of Contract Particulars

In addition to the occurrence of events during the course of the works that may give rise to a claim, it should be the duty of one of the management team, probably the contractor's quantity surveyor, to check if any fundamental changes have occurred between the contract documentation and the works themselves. Some examples will serve to illustrate this aspect:

(1) the site area and accesses, which may be more restricted than was indicated in the contract documents;
(2) the substructure, as the details may be incomplete at the design stage;
(3) the proportions and quantities of each type of work in the contract bills — it is just possible that $700\,000$ m^3 of excavation has been incorrectly billed as $70\,000$ m^3 ; and

(4) the nature and character of the works, to determine whether any significant change has occurred between the project as construed from the contract documents and that subsequently required on the site.

A check should also be made of the effect of variations on:

(1) the contract programme and period; and
(2) the smooth running of the project.

When checking any such fundamental differences or on-site problems, it is important that all relevant information regarding the assumptions made by the estimator at the time of tender is made available. For example, the length of time that earthwork support items and plant should be on the site, and the assumptions made about winter and summer working should be disclosed.

The contractor must be on the look-out for any discrepancies or divergences between any of the contract documents, such as substantial differences between scaled and figured dimensions on drawings, incorrect concrete mixes and areas of fabric reinforcement. Inconsistencies sometimes occur between the quantities of excavation on the one hand and fill and remove from site on the other. Changes in design do not always find their way through to the contract drawings and bill of quantities.

The character of varied work may be so different from the original that the billed rates are no longer applicable, as for example where a decision is made to incorporate a bridge or underpass in a roadworks contract. A substantial variation in the conditions under which the work is to be carried out will also necessitate varied rates such as the substitution of restricted access to the site for free access, curved work in lieu of straight work or the need for manual work in place of mechanical work.[2]

Armstrong[12] has described how making a loss does not necessarily imply that the contractor has grounds for making a claim. When profits are lower than expected, the contractor should examine and compare his progress against the original programme. Some of the associated problems may be of his own making or responsibility, whereas others could emanate from actions of the employer or engineer. Responsibility can normally be identifiable from the contract, although there are likely to be some borderline cases. Aspects usually attributable to the contractor may, in fact, be the employer's responsibility since his actions or those of his representatives may have caused the contractor to incur extra cost.

It is often difficult to determine where the extra cost has arisen but little progress can be made until this is resolved. However, once identified, further scrutiny will indicate whether the contractor is entitled to reimbursement and whether a claim or claims should be submitted.

The progress plotted against the programme will often identify delays, out of sequence working, disruption and disorganisation. Examination of the cause may result in a justifiable claim being made. Similarly, a comparison of expenditure on labour compared with output, may show excess cost, and this may also apply to plant and preliminaries, and the reasons for this need identifying.

If the contract is running late an extension of time for completion will often be requested by the contractor, and this requires careful scrutiny by the engineer. Some extensions of time carry money entitlement whereas others do not.

PREPARATION OF CLAIMS

Selection of Appropriate Contract Clauses and Giving of Notice

It is important that the contractor selects the relevant clause in the ICE Conditions as the basis for his claim. If in doubt it is better to select more than one clause and leave the final decision to the engineer. He also has to consider the contractual implications of the choice and may wish to have some measure of flexibility. In selecting the clauses to use, the contractor should refer to past judgements and appropriate works of reference such as Hudson[13] and Abrahamson.[14]

The period for giving notice may affect the choice of the clause. Where the timing is specified, the contractor must comply with it, otherwise the entitlement may be lost or reduced. Alternatively, the contractor may have to use another clause as the basis for his claim. The engineer may have power to make an award from the information subsequently brought to his attention, when the notice has been given late or not at all. In these circumstances the award may be assessed on the low side.

A brief memorandum incorporating the notice is normally sufficient, fuller details being provided later. The briefer the notice the better, since the contractor may have only part of the information available at this stage. Notifications must be in writing in accordance with the contract clauses selected.

Failure to notify a claim will often compel a contractor to claim his entitlement under a less favourable clause. Insufficient records may restrict his choice or force him to change later. Good records, verified by the resident engineer, are vitally important and must be retained after completion of the contract until all matters have been settled.[12]

Once the principle of the claim has been accepted by the engineer, the contractor should present a fully quantified claim. This is still necessary even where the engineer does not accept the claim in principle and it is the intention of the contractor to seek a formal decision of the engineer under clause 66 of the ICE Conditions. Should the contractor be dissatisfied with the decision, the dispute can be referred to arbitration, when the fully prepared quantified claim will form part of the points of claim prepared for the arbitration.[11]

Supporting Information

Claims are submitted principally to cover the extra cost and/or expense resulting from disruption of the work or prolongation of the contract. In either case a substantial amount of supporting information is needed in order to prepare a sound and logical claim. One useful approach is for the contractor to require site agents and foremen to insert daily comments against a numbered list of topics to avoid significant omissions. It is also wise to keep two sets of important records in different offices to guard against possible loss in the event of fire.[7]

In compiling a claim, a contractor may need to refer to any of the following documents:

(1) contract correspondence (seems straightforward at first sight but it can have important implications);

(2) approved minutes of head office and site meetings (can contain instructions, variations and additional requirements);

(3) engineer's instructions, variation orders and site instructions (could be the most important single item);

(4) contract and working drawings, including all revisions and bar bending schedules, and other contract documents (identify divergences and inconsistencies between them);

(5) labour allocation sheets (showing location, tasks and standing time);

(6) correspondence with and claims from sub-contractors and suppliers (may indicate additional requirements from main contractor or causes of delay);

(7) site diary (must contain accurate and comprehensive entries and will often highlight problems);

(8) daily weather reports (rainfall records for the previous 5 years are useful — a CIOB analysis[15] shows rain occupying 4 to 7 per cent of daytime hours with considerable monthly variations; topographical features can cause variations within a few kilometres);

(9) receipt of drawings schedule (identification of revisions to drawings as compared with those on which tender was based);

(10) progress photographs, dated by the photographer (can show lack of progress and identify disruptions as basis for prolongation claims);

(11) site survey and level details (basis for earthwork quantities);

(12) site records showing delays and disturbances (could be an important element in substantiating claims for loss and expense);

(13) photographs and report detailing condition of site at date of possession (could show obstructions and part only of site available);

(14) records showing time period between date of tender and date of possession, or order to start work (could show delay at outset);

(15) build up of tender (particularly allocations for preliminaries and general items, site and general overheads and profit);

(16) extension of time claims and allowances certified by engineer;

(17) measurement files and interim certificate measurements;

(18) materials schedule (quantities received and delivery dates);
(19) invoice lists (additional costs under fluctuations clause, where appropriate);
(20) plant records (showing standing time and number of times brought to and from site);
(21) statutory undertakers' records;
(22) site costs and finance (showing how money has been spent and extent of reimbursement);
(23) authorised daywork schedule (covering varied work which cannot be valued at billed rates);
(24) programmes and progress charts (showing contractor's anticipated programme and actual performance);
(25) borehole logs (actual soil conditions may be different from those which the contractor could reasonably have anticipated);
(26) work method study (identifies extent to which disruption has occurred and its effect);
(27) variation data sheets (nature and effect of variations);
(28) interim applications, certificates and payments (amounts and pattern of payments — cash flow aspects); and
(29) schedules of defects.

Possible Consequences of Engineer's Instruction

It is relatively easy to find correspondence on the files that confirms a major instruction, but rarely does it amplify the probable consequences which are so important when preparing a claim. For example, the following consequences may result from an engineer's instruction:

(1) work in several areas will be disrupted;
(2) labour and plant will have to be transferred to other areas;
(3) new material will have to be ordered, and replaced materials either scrapped or transported to another site;
(4) other related work will be more costly;
(5) delay will be caused and extension of time will be necessary; and
(6) reimbursement of loss and expense will be sought.[3]

Significant Documents

One of the most important documents is the labour allocation sheet showing who is working where and what they are doing, with space for entries relating to disruption. Site management need to know and trace all periods of work that are unproductive. These must be shown on allocation sheets together with the reasons for stoppages. They may result from the issue of engineer's instructions or shortcomings within the contractor's own organisation. A knowledge of both is vital to management.

Another important record is a variation data sheet which asks a whole range of pertinent questions, including the following:

(1) Was there a stoppage of work by operatives?
(2) Were any operatives redirected?
(3) Was there any loss of output of plant?
(4) Was any plant redirected?
(5) Were any sub-contractors affected or disrupted?
(6) Did the variation result in the order of new materials?
(7) Were any materials made surplus?
(8) How much extra staff time was incurred?[3]

The details from a variation data sheet enable a comprehensive cost assessment of the loss situation to be made, and the effects of a variation to be priced out in bill form. To secure official agreement it is advisable to request the resident engineer to examine the variation data sheet and subsequently to sign it.

Essential Requirements in Claim Preparation

One of the main criteria in establishing the validity of claims is good, accurate records. Probably the next most important step is to inform the engineer that claim situations are arising. A major problem can be the confidentiality aspect of much of the cost information which the contractor often guards jealously and some of which may be needed to satisfy the engineer of the validity of the claim.

Every claim should be produced as if it is to become evidence in court (which it may do) and should be carefully detailed and presented, preferably in a bound cover. An untidy and carelessly prepared claim is unlikely to receive very serious consideration.

The claim could conveniently be broken down into the following logical sequence:

(1) Contract particulars — details of the site (as contained in the preliminaries and general items) and details of the contract (as contained in the form of agreement and appendix to the form of tender).
(2) Claim particulars — a summary of the bases or heads of claim, stating all facts and details, together with full particulars of the specific contract clauses on which the claim is based.
(3) Evaluation of the claim — a summary of the contractor's financial loss and/or expense.
(4) Appendices — a section that collates all the back-up information described in (2) and (3).

Often the claim is based on a reasoned argument of loss rather than the proof of the loss. The cause of this may stem primarily from the variability in the quality and preciseness of the contractor's procedures. As a result of negotiation and compromise the less well-prepared claims are frequently amicably settled or withdrawn, but it is far better for the contractor to start from a sound and realistically based claim. It is also beneficial for the contractor to reach a reasonable settlement with the employer's advisers, by means of a well-presented and carefully considered claim, than to resort to costly arbitration.

To sum up, in preparing a claim the first essentials are for the contractor to determine the extent of his obligations under the contract and then to obtain details of the matters that hindered or prevented him from executing the work with expedition and economy. It then remains for the facts to be stated with the utmost precision and clarity and to calculate the amount of the additional expenses incurred. It is bad policy for the contractor to submit inflated claims using the argument that he does not expect to receive the whole of his entitlement. It is far better if realistic claims are submitted and a truly professional attitude is adopted by all the disciplines concerned.

ACTION ON SUBMISSION OF CLAIMS

Control of Claims

The only certain way of avoiding the payment of contractors' claims is to eliminate the issue of variations and the disruption to regular progress that accompanies them, and to operate the contract so efficiently that the contractor will not incur additional loss and/or expense, but this ideal situation is rarely if ever achievable. Trickey[8] has described how a contractor cannot be prevented from submitting a claim, no matter how ill-founded. However, the submission of a claim by the contractor does not automatically create a right to reimbursement. The contractor must first demonstrate that a real loss or expense has been caused directly by one of the criteria recognised in the contract. On the other hand, the employer should not avoid making proper recompense to the contractor should the engineer issue instructions varying the work or otherwise disrupting the contractor's progress.

The only way to properly assess the sum due to the contractor is for the engineer to be notified when the event occurs that gives rise to the claim. Allegations can then be examined against the prevailing facts, and the real impact of any delay, disruption and the like can be fully investigated. Uncertainty creates the wrong environment in which to expect even an efficient organisation to make satisfactory decisions. When variations are issued, the value both in time and money should be quickly assessed so that the probable financial effect is monitored and controlled.[8]

On occasions, claims are successfully pursued which need not have arisen, while others are settled at unrealistic levels when considering all the relevant facts. These situations cannot be in the best long-term interests of the construction industry and highlight the need for an impartial and effective treatment of claims.

The number of claims can be reduced by the preparation of accurate and complete contract documents. The design team should take particular care to:

(1) eliminate discrepancies;
(2) resolve details of all nominations; and
(3) co-ordinate the work of any other contractors who are directly appointed by the employer.

The engineer and/or quantity surveyor should have full regard to the following elements when assessing the contractor's entitlement, namely: materials, labour disruption, attraction money and bonus payments; preliminaries, general items and plant; inflation; head office overheads and profit; and interest charges.[8]

Basis of a Successful Claim

Armstrong[12] has described how many claims are based on the hope that the engineer's sympathy will suffice. There may be grounds for an *ex gratia* payment but these are rarely made, particularly in public sector contracts. An entitlement within the contract must be shown, stated and proved, and it is necessary to give the clause(s) under which the claim is made. Most contract clauses require the employer's responsibility to be clearly demonstrated , and this includes action or inaction of the engineer and others.

Where the terms of the contract require notification of a claim, and most of them do, it is essential that this is given in accordance with the appropriate time limits. Failure to do this may not necessarily invalidate the claim but, as described earlier in the chapter, the cost and expense is more likely to be disputed and reduced to the contractor's disadvantage.

An extension of time must be requested where appropriate. It may be needed to reduce liquidated damages as, for example, in the case of adverse weather conditions where additional cost and expense are normally the responsibility of the contractor. Equally, an extension may form the basis for a claim for entitlement to increased preliminaries and overheads, or extra labour and plant. The contractor should ensure that he is not granted excessive extensions of time on items which carry no money entitlement, as these may prejudice his subsequent claims for reimbursement. An extension of time does not always lead naturally to a claim for entitlement. The claim may be for an item which is not on the critical path, or the time may be regained at extra cost and expense.

Often when the engineer is notified of a claim he is required to state what records he requires the contractor to keep. The contractor must, in any event, maintain adequate records to establish the extra cost and expense incurred. Armstrong[12] has described how many claims fail because the records are inadequate or incorrect. This may result in the engineer awarding an amount less than the sum to which the contractor is entitled. Records must be accurate and consistent, substantiated by the resident engineer, and often supported by photographs.

Action by Engineer pursuant to Receipt of Notification from Contractor

Where a contractor gives notification of his intention to submit a claim under sub-clause 12(1) of the ICE Conditions for additional costs incurred in dealing with physical conditions and/or artificial obstructions, which he could not reasonably have foreseen, the engineer under sub-clause 12(2) may if he thinks fit:

(1) require the contractor to provide an estimate of the cost of the measures he is taking or proposes to take;
(2) approve in writing such measures with or without modification;
(3) give written instructions as to the method of dealing with the physical conditions or artificial obstructions; or
(4) order a suspension under clause 40 or a variation under clause 51.

Should the engineer decide that the claim is acceptable in whole or in part, then under sub-clause 12(3) he is required to:

(1) determine any extension of time for completion under clause 44;
(2) certify payment of reasonable costs for additional work done;
(3) certify payment of reasonable costs for additional constructional plant used;
(4) certify payment of reasonable costs incurred by the contractor by reason of any unavoidable delay or disruption of working suffered by him; and
(5) certify payment of a reasonable percentage addition in respect of costs of additional work done and plant used, but not on delay and disruption costs.[11]

Rate Fixing

Rate fixing normally involves two factors:

(1) agreement to a star item which consists of a new major item of measured work which has no comparable description in the bill of quantities, and the fixing of a rate known as a star rate; and
(2) fixing rates for new subsidiary work items arising out of a star item.

The contractor should write to the engineer requesting star rates before commencing work on star items or as soon after as is reasonably practicable.

New rates become necessary through a variety of factors of which the following embrace some of the more commonly encountered ones:

(1) variation(s);

(2) changes or modifications in design;

(3) new or additional work;

(4) result of omissions;

(5) alterations or substituted work;

(6) delay or disturbance caused by others;

(7) deviation in quality or type of labour and materials;

(8) effect of quantities which differ widely from those contained in the original bill of quantities;

(9) ambiguities and discrepancies or divergences or errors in the bill of quantities or contract drawings;

(10) uneconomic working;

(11) standing or idle time of labour and of plant;

(12) work carried out under different conditions from the original contract requirements;

(13) work of a different character from that contained in the bill of quantities;

(14) execution of expedient or emergency work;

(15) night and/or weekend work;

(16) progress and rate of work retarded because of special directions;

(17) acceleration measures;

(18) work below groundwater level;

(19) work between tides;

(20) work within cofferdams;

(21) work in running sand or water;

(22) work at widely differing heights;

(23) late instructions; and

(24) work out of sequence.[16]

ADDITIONAL COST FOR LOSS OR EXPENSE

Background to Claims

Direct loss and/or expense often results from disturbance of the regular progress of the works and this may arise because of lack of instructions, drawings, details or levels; the opening up for inspection or testing if found unnecessary; discrepancy between contract drawings and bill of quantities; delay on the part of other contractors employed by the employer; and engineer's instructions covering postponement of any work under the contract.

If the contractor follows the correct procedure with regard to notification and the engineer is of the opinion that the contractor has been involved in direct loss and/or expense then the engineer will ascertain, possibly in conjunction with the quantity surveyor, the amount of such loss or expense. This ascertained loss is added to the contract sum and included in an interim certificate if one is to be issued after the date of ascertainment.

The probable causes of interruption can also become grounds for extension of time (sub-clause 44(1)), as they are under the control of the employer or engineer. There are also matters for which extension of time may be granted which would not necessarily provide grounds for a monetary claim, such as *force majeure*; exceptionally adverse weather; strikes and lock-outs; and delays in securing labour or materials.

The engineer is empowered to issue further drawings and instructions from time to time during the course of the work. The timing of requests by the contractor to the engineer for instructions is significant and it must be realistic. For example, it would be unreasonable for a contractor to request a bar bending schedule in the second week of a contract for reinforced concrete work to be executed in the thirty-third week and then to regard the immediate non-provision as a basis for claim. This contention was supported in *Neodox Ltd v. Swinton and Pendlebury* BC (1958)5 BLR 34, wherein it was held that the contractor cannot expect all information on day one. It was further held that such details or instructions should be given within a time reasonable in all the circumstances — including the convenience of the engineer.[6]

Accurate daily or weekly reports to the engineer, together with an agreed programme of work, preferably supported by a critical path network and a work method statement, should provide a realistic forecast of when information is reasonably required. This can be further refined by the supply by the contractor of key dates when specific details and information will be required.

Contractors' claims embrace a wide range of matters and the following list serves to illustrate their diversity:

(1) Cost of bringing to the site, plant and operatives whose work was finished, and their subsequent transfer to other sites.
(2) Cost of extra hire periods of plant.
(3) Cost of delay in the non-use of plant.
(4) Extra cost of moving unwanted or unused plant.
(5) Cost of waste of materials, such as cement and lime, through deterioration caused by long delays. The contractor must, however, provide adequate storage facilities and use the materials in the proper sequence.
(6) Cost of uneconomic use of plant and labour when diverted to other work, to avoid waiting or standing time.
(7) Cost of extra power, lighting and watching.
(8) Cost of extra de-watering, as a result of the extension of underground working.
(9) Cost of ready mixed concrete instead of site mixed concrete, after the removal of the batching plant.
(10) Cost of labour operations as against mechanised operations, after the removal of plant.[17]

Changes in the engineer's requirements may have significant financial effects. For example, substituting 300 mm reinforced concrete walls for 150 mm walls will have a substantial effect on formwork prices, although on occasions, engineers and quantity surveyors have attempted to apply the same billed rates. Formwork to a 150 mm wall might, for instance, use 12 mm plywood with 50 × 50 mm softwood soldiers at 450 mm centres and Acrow propping at 900 mm centres. The 300 mm wall will require much stronger formwork, possibly 19 mm plywood and much stronger soldiers and closer propping.[17]

Claims may be made, for example, under clauses 12, 13 and 14, relating to adverse physical conditions and artificial obstructions, changed mode and manner of construction, and varied constructional methods, if the regular progress of the work is materially affected. This can happen without needing to extend the contract period. It may be that the varied work is not a critical item and can be undertaken in float time on the critical path network. It is disturbance that costs money even although the work in its entirety is not delayed. Such claims may revolve around acceleration with the need for increased productivity of labour or plant, the necessity of providing additional supervision and other related matters which are not time-related.

The late supply of drawings or other information by the engineer could result in the delayed delivery of materials which disrupt the work. The contractor may then have to authorise overtime working to keep the work on target and so incur additional cost which will be recoverable from the employer.

The tender for a project is based upon work being carried out in an orderly sequence with the most efficient use of labour and plant. A disturbance to flow through lack of details could conceivably increase the cost of the work by as much as 50 per cent, as the rate of output could revert from the peak rate to the much lower rate prevailing at the start of the contract. It must be borne in mind that other factors can also reduce productivity rates, such as exceptionally adverse weather, go-slow tactics by labour, poor supervision, inefficient subcontractors and poor or insufficient labour.

Assessment of Claims

In order to assess a claim the engineer will often have to carry out extensive investigations. For instance, when faced with a claim concerning the additional cost of executing work brought about by delay in the supply of certain drawings, he would need to:

(1) ascertain the period between the latest date when drawings were required and the date at which they were actually supplied to the contractor;
(2) ascertain on which work the men were employed who would otherwise have been engaged on the work detailed on the missing drawings;
(3) determine whether they were transferred to this latter work immediately the drawings were supplied; and
(4) ascertain the overall programme and labour position at the time the arrival of the drawings was being awaited.

The claim may be based on additional cost due to operatives standing idle or to indirect problems such as the reprogramming of the works with consequent inconvenience and loss of efficiency.[2]

Another possible source of claims is where several contractors are working on the same site at the same time, as on a power station site, where building and civil engineering, mechanical and electrical contracts are in operation, with a distinct possibility of interference between the contractors, despite the careful wording of the contracts and excellent co-ordination arrangements on site. The superstructure contractor may be delayed while certain plant is being installed, then he has to leave access gaps in the structure and subsequently to close them at a much later date, possibly beyond the end of his contract period.

As the duty of the engineer is to ascertain the amount of the direct loss and/ or expense, it is necessary for him to look to the contractor to supply all necessary factual information. Sub-clause 52(4) of the ICE Conditions provides that the contractor must submit on request such details of the loss and/or expense as are reasonably required by the engineer. This may not necessarily mean the submission of a fully costed claim, although it is usual for the contractor to supply it in accordance with sub-clause 52(4)(d). Powell-Smith and Sims[18] suggest that on building contracts the details might include such aspects as comparative programme/progress charts pin-pointing the effect upon progress, together with the relevant extracts from wage sheets, invoices for plant hire and other related documents.

When evaluating loss and expense, the engineer has always to contrast what has happened with what would have happened had not the delay or disruption occurred.[5] Almost inevitably an element of conjecture is involved but the engineer, possibly assisted by a consultant quantity surveyor, has throughout a professional duty to assess the matter in a fair and impartial manner and to exercise due professional skill and judgement. The construction industry in the United Kingdom is generally believed to be potentially the most efficient in the world, but it does suffer from the large number of disputes that arise, relating principally to direct loss and expense.

DISRUPTION OF THE WORKS RESULTING FROM VARIATIONS

The issue of certain types of variation order can have a significant effect on the efficiency of a project, and its resultant productivity and profitability. For example, a variation encompassing an extension to a sewage treatment works under construction which impinges upon the contractor's main access to the site, will have a disruptive effect on the movement of operatives, plant and materials on and off the site. Similarly a variation changing drainage pipes from unplasticised PVC to clay could affect the overall site planning. Proving that such disruption falls within the generally accepted criterion of 'direct loss and/or expense' can be difficult. Although there are some fundamental principles that are important in the preparation and assessment of claims, each case must normally be treated on its merits.[2]

Programme

A realistic, properly prepared and detailed programme is a vital aid to proving disruption of the programme, particularly when it identifies critical activities.

Resource Allocation

As an aid to proving the costs of disruption it is advisable for the contractor to show, in a simple form on the programme, the resources he has planned to use throughout its various stages; for example, the number of operatives to be employed per day or per week, and the amount and type of plant expected to be used. These can then be compared with the records of the actual resources used on the works and the cost differential established. The resident engineer and his staff normally keep their own records which can be used as an independent check. The type of information described may also be incorporated by the contractor in a method statement.

Effect of Variations

The disruption or adverse effect of variations should be suitably recorded at the time the variation orders are issued. These are best entered on a standardised form containing the type of questions posed earlier in the chapter for insertion on a variation data sheet. The information required in response to the questions will be inserted by site staff, as a basis for a claim to be submitted by management staff.

Acceleration

Acceleration entails increasing the rate of the constructional work to meet the contract completion date. This can be achieved by increasing the number of operatives on site, the length of working hours or the amount of plant. If the issue of an engineer's instruction has caused delay it may be in the best interests of all concerned to make up the time lost by acceleration. It would, however, be unwise for the contractor to expend money in this way without first obtaining the engineer's agreement and his confirmation that he is prepared to meet the cost.

Since the alternative to this kind of disruption claim is likely to be a request from the contractor for an extension of time under sub-clause 44(1) of the ICE Conditions, with an associated request for recompense under sub-clause 52(4) (a prolongation claim), the engineer will usually decide to accept the cost of acceleration, as this is likely to cost less than the other available alternatives and will ensure that the project is still completed on time. The cost of such an acceleration claim is normally obtained by using programme comparisons, as previously described. Revised programmes, but incorporating the same details, should be supplied by the contractor to the engineer.

Recovery of the Cost of Variations

Contractors have in the past often co-operated with engineers and quantity surveyors in applying *pro rata* rates as a ready means of valuing varied work. This procedure should, however, be applied only when the work is of similar character and executed under similar conditions to the work priced in the bill of quantities. In practice there are few variations that can be made to the contract works without altering either the character and/or the conditions under which the work will be performed, and in these circumstances it is necessary to assess fair rates and prices.

Final Accounts

Some engineers appear to treat a final account as a technical operation rather than a matter of judgement. In some cases a project is subject to such extensive redesign that it bears little resemblance to the original scheme. The engineer and/or quantity surveyor may decide to abandon the preparation of a bill of variations, as used extensively on building contracts, and to remeasure the whole of the work, including the varied work, at billed rates. This is tantamount to an admission that the entire contract is varied and it is then patently incorrect to use a bill of quantities in the same way as a schedule of rates, and new rates and prices should be negotiated to arrive at a fair valuation of the work.[4]

Prolongation Claims

These claims may be presented as a result of extensive variations and be associated with a claim for extension of time under sub-clause 44(1), or as a result of interference with the regular progress of the work under sub-clause 52(4) of the ICE Conditions. Prolongation claims will be based on the loss and/or expense incurred by the contractor and resulting from the extension of the contract period. The various elements likely to appear in this type of claim are now separately considered.

Preliminaries and General Items

Contractors often claim an adjustment of preliminaries and general items because an extension of time has been granted. These claims are only admissible where the delay is attributable to an action of the employer, or has resulted from the effect of adverse physical conditions and artificial obstructions described in clause 12 of the ICE Conditions. When the claim is admissible, only reasonable additional costs which can be verified by the engineer should be submitted.[19]

The value of preliminaries and general items may be increased because of a number of the components being directly related to the contract period. A claim under this head should preferably be based on actual expense, and not merely to *pro rata* the prices inserted against preliminaries items. The contractor should record the actual expense incurred during the period of extension. Another approach is to sub-divide the preliminaries items into the three categories of lump sums related to specific events, time based and value based. The main expenses involved are likely to include the following:

(1) salaries of site staff (time based);
(2) cost of plant and vehicles retained on site (time based);
(3) temporary lighting (time based but adjusted for the time of the year);
(4) offices and stores retained on the site and their upkeep (time based but with allowance made for erection, dismantling and transport);
(5) safety measures (time based);
(6) protection of the works (time based);
(7) insurance premiums (value based);
(8) telephone costs (specific event);
(9) electricity costs (specific event); and
(10) rates on site buildings (specific event).

Time of Year when Work is Carried Out

The extension of a contract from summer into winter may lead to extra costs as a result of reduced working hours, stoppages through bad weather and other associated additional costs. Carefully kept site records will clearly identify these additional costs which can then be compared with the programme and resource allocation.

Extended Attendances on Nominated Sub-contractors

When the contractor prices attendance on nominated sub-contractors these may, in certain circumstances, relate to the period of time that the nominated sub-contractor is on the site. When as a result of variations or extensions of time these attendances are extended, the contractor should record the actual costs incurred.

Overheads

Claims should ideally reflect actual costs or expenses that are provable and this principle should, as far as practicable, be applied to site overheads. In the case of general overheads, it can be argued that the mere extension of a contract will not automatically result in increased costs. If it has been necessary to increase the number of office staff employed as a direct result of the prolongation then the cost of such an increase would be claimable, based on the actual additional cost incurred. Otherwise it could be argued that the overheads are the same but that their incidence is spread over a longer period.

Overheads are a budgeted amount calculated on the anticipated year's expenditure on staff, offices, equipment, stationery and the like, and are related to a budgeted year's turnover. On this basis a minimum percentage on a fixed turnover must be achieved to recover costs. If the turnover is significantly reduced then the percentage of overheads needs increasing to produce the required sum. If the contract period is extended, work of the same value is spread over a longer period and, as a result, turnover in a given year is reduced.

To recover his overheads the contractor may need to increase his overheads percentage, but on current work the tenders will have already been submitted and accepted, so it is too late to adjust them. This could form the basis for an under-recovery claim as illustrated in table 8.1, and this represents one way in which an overheads shortfall could possibly be substantiated.

Loss of Profit

Most contractors' claims will include some element of profit. However, a claim of this kind is very difficult to prove, since profit can be lost by inefficiency or bad tendering before any extension of time occurs. There are various schools of thought:

(1) It can be argued that an overall shortfall of £400 000 (as calculated in table 8.1) will result in a profit shortfall in twelve months of say 3 per cent of £400 000 = £12 000.
(2) Another argument is that since the total value of the work was not reduced, although it was spread over a longer period, neither was the profit reduced; thus there was no actual loss.
(3) Yet another approach could be that keeping key operatives on the site for a longer period than anticipated reduces their profit earning capacity elsewhere.

The definition that loss is something the contractor should have received but did not is untenable. It could possibly be argued that in case (1) the company will not be able to pay the desired return to its investors on their invested capital as a direct result of the anticipated shortfall arising from the extension of the contract. The most common result of a loss of profit claim is a compromise.

In disruption claims the biggest problem for all parties lies in the interpretation of the word 'direct' which is normally applied to loss and expense. In modern high-speed construction, the disruption of one trade or area of work reverberates through all following trades and, if the contract is bedevilled with disruptive incidents, the tracing of individual losses is virtually impossible.

Table 8.1 Under-recovery claim

Budgeted turnover in 12 months	£13 000 000
Estimated cost of general overheads	£780 000 (6 per cent)

Assume that a contract valued at £1 600 000 is extended three months, where the contract was expected to be completed in twelve months. The possible value of the shortfall could be £1 600 000 × 3/12 = £400 000, being the value of the work that could have been done in the extended three months period.

Therefore the turnover in this particular year would be reduced in the manner shown:

budgeted annual turnover	£13 000 000
less calculated shortfall	400 000
actual turnover	£12 600 000

When tendering, general overheads are expressed as a percentage of the estimate, and once a project is obtained there is no way of increasing this percentage. Hence the overheads recovery could be calculated as:

6 per cent of £12 600 000 = £756 000

This shows an under-recovery of

£780 000 – £756 000   = £24 000

Claims are then made by the comparison of tender totals of operatives and plant hours with the actual hours expended. This tends to reduce the contract to a prime cost situation and makes no allowance for the possibility of poor planning, mistakes, bad workmanship and dilatory management by the contractor.[3] These problems highlight the difficulties inherent in the fair and reasonable assessment of claims of this type.

LIQUIDATED DAMAGES

The ICE Conditions of Contract (clause 47) provide for the payment of liquidated damages for delay in completion beyond the completion date inserted in the appendix to the contract, or a substituted date following the grant of extension of time by the engineer. The sum inserted will be the amount payable for each day, week or other prescribed period for which completion is delayed. A sum inserted for liquidated damages to be enforceable must be a genuine pre-estimate of damages, and should represent the likely financial loss or cost incurred by the employer if delay occurs. If the amount is not a genuine pre-estimate of damage it could be held by the courts to be a penalty. In these circumstances, the employer can only recover his actual loss and not the amount of the penalty. Whether the sum is a penalty or is liquidated damages will be largely influenced by the terms and inherent circumstances. In many civil engineering contracts the liquidated damages are related to those included in previous contracts of a similar nature.

The Society of Chief Quantity Surveyors in Local Government (SCQSLG) set up a working party to investigate the procedures adopted for the assessment of liquidated damages on local authority building contracts.[20] The working party viewed the wide range of damages used with some disquiet but found no individual rates penal. The investigations found that case law gave little help on formulating precise methods of assessment and there was no evidence of damages ever having been set aside by a court as a penalty.

The precedent set by the courts for a valid assessment of damages was summarised as follows:

(1) If the parties made a genuine attempt to pre-estimate the loss likely to be suffered, the sum stated will be liquidated damages and not a penalty, irrespective of actual loss.
(2) The sum will be a penalty if the amount is extravagant having regard to the greatest possible loss that could be caused by the breach.

The form the loss might take was illustrated by the working party[20] by taking extracts from *Halsbury's Laws of England* (Third Edition):

(a) 'The measure of damages for failure by the contractor to complete a building . . . will include . . . any loss of rent of the building, or any loss of use of the building or, in appropriate circumstances, of business profit which may accrue to the employer in consequence of any delay . . . '
(b) 'In certain cases the measures of damages may be the loss of interest on the cost of the contract works and of the land on which they are constructed'.

The SCQSLG working party[20] examined the alternative approaches and concluded that loss of interest on the cost of the contract works constituted the only direct loss that was legal and capable of genuine pre-estimation with a minimum of problems. After considering responses from 53 local authorities, the working party recommended that the following approach provided a reasonable basis for the assessment of liquidated damages:

(1) Loss of interest on the cost of the contract works, based on the assumption that 80 per cent of the monies is paid at theoretical completion and that annual interest payable is 16 per cent. This amounts to 0.2 per cent of the

contract sum per week but will need monitoring to take account of significant fluctuations in interest rates.

(2) Professional fees of the architect/engineer and site staff, calculated as a percentage of the contract sum.

(3) Further costs, such as costs of a temporary nature awaiting the completion of the project, such as temporary housing, and additional costs of the employer which are best assessed on an *ad hoc* basis.

(4) Fluctuations, where applicable, normally calculated by reference to the index numbers relating to the valuation period.

Marks *et al*[19] have described how liquidated damages inserted in the appendix to a contract for the construction of a wharf or jetty could be related to the expected loss of profit of the owner for each day that he is deprived of possession of the new wharf or jetty. It is intended to be a reasonable pre-estimate and cannot be varied if for any reason the forecast is inaccurate.

It will be more difficult to assess the rate of liquidated damages for work which on completion will not be a source of direct gain to the promoter, such as the construction of a motorway. The delay in completion of a sea wall by three months may only result in loss of amenity, but the situation would be entirely different if the contract extended into the winter and an exceptionally high tide occurred during the period of delay and caused extensive damage to adjoining property.

Although the estimate of liquidated damages is not expected to be precise, it must be reasonable. Thus the inclusion of damages of £2500 per week in respect of a pumping station costing £76 000 to construct would be considered grossly excessive.[19]

Substantial completion of the contract will normally release the contractor from his obligation to pay liquidated damages, particularly where the employer has taken possession of the whole or part of the structure. An employer could not claim liquidated damages in respect of non-completion of a power station, merely because an ancillary building, which would not prevent the power station being operated, was incomplete. On the other hand, where statutory consents are required, the project is incomplete and cannot be used, and liquidated damages for delay are payable until the necessary consents have been obtained.[19]

Chappell[9] has listed the following advantages of liquidated damages:

(1) they do not have to be proved;

(2) they are agreed between the parties in that they are known to the contractor at the time of tender and he can allow for them in his tender total; and

(3) the employer can simply deduct them without having to issue a writ through the courts.

REFERENCES

1. Institution of Civil Engineers, Association of Consulting Engineers and Federation of Civil Engineering Contractors. *Conditions of Contract for use in connection with Works of Civil Engineering Construction.* Fifth Edition (June 1973, revised January 1979)
2. I.H. Seeley. *Quantity Surveying Practice.* Macmillan (1984)
3. T.H. Robinson. *Establishing the Validity of Contractual Claims.* Chartered Institute of Building (1977)
4. H.T. Burke. *Claims and the Standard Form of Building Contract.* Chartered Institute of Building (1976)
5. G. Trickey. Evaluating contractors' claims: the professional quantity surveyor's approach. *Chartered Quantity Surveyor* (February 1979)
6. G.A. Hughes. *Building and Civil Engineering Claims in Perspective.* Construction Press (1983)
7. T. Davies, H. Hay and J. Sneden. Processing civil engineering claims. *Chartered Quantity Surveyor* (November 1980)
8. G. Trickey. *The Presentation and Settlement of Contractors' Claims.* Spon (1983)
9. D. Chappell. *Contractor's Claims.* Architectural Press (1984)
10. D.T. Simmonds. Evaluating contractors' claims: presentation of claims by contractors. *Chartered Quantity Surveyor* (February 1979)
11. C.K. Haswell and D.S. de Silva. *Civil Engineering Contracts: Practice and Procedure.* Butterworths (1982)
12. W.E.I. Armstrong. *Contractual Claims under the ICE Conditions of Contract.* Chartered Institute of Building (1977)
13. I.N. Duncan Wallace (Editor). *Hudson's Building and Engineering Contracts.* Sweet and Maxwell (1979)
14. M. Abrahamson. *Engineering Law and the ICE Conditions.* Applied Science Publishers (1979)
15. Chartered Institute of Building. Site management information service nr 87. *Weather and Construction* (1981)
16. R.D. Wood. *Building and Civil Engineering Claims.* Estates Gazette (1985)
17. R.D. Wood. *Builder's Claims under the JCT Form of Contract.* Chartered Institute of Building (1978)
18. V. Powell-Smith and J. Sims. *Building Contract Claims.* Granada (1983)
19. R.J. Marks, R.J.E. Marks and R.E. Jackson. *Aspects of Civil Engineering Contract Procedure.* Pergamon (1985)
20. Society of Chief Quantity Surveyors in Local Government. *Assessment of Liquidated and Ascertained Damages on Building Contracts* (1981)

Index